Dragons and Damsels

An Identification Guide to the British and Irish Odonata

Adrian M. Riley

Brambleby Books

Dragons and Damsels – An Identification Guide to the British and Irish Odonata
Text © Adrian M Riley 2020

The author has asserted his rights under the Copyright, Designs and Patents Act 1988 to be identified as the Author of this Work. All Rights Reserved.

No part of this book may be reproduced in any form by photocopying or by any electronic or mechanical means, including information, storage or retrieval systems, without permission in writing from both the copyright owners and the publisher of this book.

ISBN 9781908241641

Published 2020 by
Brambleby Books Taunton, Somerset, UK

www.bramblebybooks.co.uk

Book layout by Tanya Warren, Creatix Design

Printed on FSC paper and bound by Cambrian Printers U.K.

**To my wife
Marian**

I'm happy; hope you're happy too.
David Bowie (1947–2016)

Scarce Chaser immature male. Strumpshaw Fen, East Norfolk (p 150).

About the Author

Adrian Riley is the author of numerous books about wildlife, including field guides such as his best-selling *British and Irish Butterflies* and *Norfolk Wildlife*, and his anthropological account *Arrivals and Rivals: duel for the winning bird* (all three published by Brambleby Books), the latter concerning his aim to see as many bird species as possible throughout the British Isles within a single year. Adrian grew up in Shropshire where his deep interest and knowledge about wildlife, especially butterflies, moths and birds, first took root.

For forty years he has worked as a professional entomologist and moth recorder for the world-famous Rothamsted agricultural research station in Harpenden, Hertfordshire. For the past 15 years or so, he has lived in Norfolk from where he has led wildlife tours, both within the county, the UK and mainland Europe.

The author in the Outer Hebrides

Emerald Damselfly

Contents

Foreword by Dr Pam Taylor	10
Acknowledgements	13
Introduction	14
How to use this book	16
Systematic Checklist and Status of all Species and described forms recorded in Britain and Ireland	18
Anatomy	21
Chapter One: Species accounts of the resident Damselflies	25
Chapter Two: Damselflies: Identification of the resident Species	64
Chapter Three: Species accounts of the resident Dragonflies	113
Chapter Four: Dragonflies: Identification of the resident Species	170
Chapter Five: Annual Visitors and probable Colonists	230
Chapter Six: A Review of scarce or assisted Visitors and extinct Residents	235
Appendix: Vice-counties of Great Britain and Ireland	238
References and further Reading	240
Index	244

Foreword

The precursors of our modern dragonflies and damselflies first took to the air over 300 million years ago. Although some of the early examples in the Carboniferous period were huge by current standards, many fossil 'dragonflies' were similar in size to those on the planet today. The largest ever early species had wing spans up to 75cm, which is four times the size of the largest species flying now. These precursors of modern dragonflies belonged to the now extinct genus *Meganeura*. Although only distantly related to those on the wing today, their appearance was very similar and certainly recognisable.

Today's largest specimens are a group of tropical damselflies, commonly known as 'Helicopter Damselflies'. They have wing spans of up to 19cm, and I have been privileged enough to see examples of these extraordinary insects in both Panama and Ecuador. The thrill as you spot these beautiful damselflies fluttering in the canopy above you can hardly be described, so how amazing would it have been to witness their huge antecedents?

Here in Britain and Ireland the largest species today are the Golden-ringed Dragonfly, *Cordulegaster boltonii*, and Emperor, *Anax imperator*, the former having the longest body while the latter has a heavier build. Both are impressive insects when seen on the wing. At the other end of the scale are the diminutive Black Darter, *Sympetrum danae*, and the delicate damselflies. This book covers the full range of British and Irish species, no matter how large or small.

I first became interested in dragonflies in 1986, and my original source of information was the field guide written by Cyril O. Hammond in 1977. My second edition copy, revised by Robert Merritt in 1983, of *The Dragonflies of Great Britain and Ireland* led me into a whole new world of interest and wonder. Of course, I'd been aware of dragonflies before, but this first field guide of mine introduced me to the myriad colours and patterns borne by the various species and sexes.

I began by studying the species local to me in Norfolk, but I was soon recording not only the ones found in my home county but also those I encountered further afield. I joined the British Dragonfly Society (BDS) early in my interest and was soon part of a nationwide 'family' of dragonfly enthusiasts. I became County Dragonfly Recorder for Norfolk in the mid-1990s and have continued my involvement with the BDS ever since.

For some unknown reason I've always been drawn to things that fly. It started with a childhood interest in birdwatching, greatly encouraged by my father who was a true countryman at heart. This interest later grew to encompass both butterflies and bats, but neither of the latter really held my attention in the way birds did, and that dragonflies would come to, when I first began to notice them. Hooked by these

Foreword

fascinating and agile aerial predators, my collection of dragonfly field guides soon grew and has continued to do so ever since.

In the main, most field guides follow the same pattern, with species accounts laid out in taxonomic sequence. The new guide by Adrian Riley does indeed have two chapters of species accounts covering damselflies and then dragonflies separately. Where this new guide differs, however, is in its approach to the identification of individual insects.

Again, damselflies and dragonflies have their own chapters, but within these, species are grouped according to their appearance, with males and females often treated separately due to their differing colours and patterns. This makes sense when you understand that early dragonfly observers actually thought that, for example, male and female Banded Demoiselles were of two different species because they looked so dissimilar.

Adrian Riley's meticulous approach to each species, sex and colour-form throughout the book should leave no-one in doubt of an identification. There is no question at all that this new guide, with its fresh approach, detailed descriptions and clear photographs, will find a place on the bookshelf of many dragonfly watchers and recorders, no matter how experienced.

Dr Pam Taylor
Trustee of the British Dragonfly Society (BDS), Convenor of the BDS Dragonfly Conservation Group and County Dragonfly Recorder for Norfolk.

Southern Migrant Hawker

White-legged Damselfly

Acknowledgements

I most sincerely thank following people and organisations for their help and support:

Roland Alderson (webcandy.com); Marian Arber; Pat Batty (BDS, Scotland); Edward Bramham-Jones (Pensthorpe NR); Steve Brooks (BMNH); www.bugalert.net; Steve Cham (BDS); Simon Davies; Andrew Dyball; Simon Lammin; Allan Livingstone; Nicola and Hugh Loxdale; Keith Mallett; Alex McLennan; Gill McLennan; Nigel Middleton (Sculthorpe Moor NR); Brian Nelson (BDS, Ireland); Adrian Parr (BDS); Fons Peels (DragonflyPix.com); Rare Bird Alert; Kerry Robinson; Ashley Saunders; Alan Shears; Graham Sherwin; Gary Thoburn; Pam Taylor (BDS); Peter Vandome; www.ukdragonflies.com.

Special mention must go to Steve Cham, Andrew Easton, Simon Lammin, Keith Mallett, Kerry Robinson, Graham Sherwin, Gary Thoburn, Paul Winter and Peter Vandrome for their generous donations of photographs.

Introduction

When I first took a serious interest in this group of insects in the 1980s, I found identifying them very frustrating. Even though I have now been an entomological taxonomist for over forty years, I still find identification of some of the damselflies challenging when working in the field.

In the mid-1980s, I was armed with Andrew McGeeney's *A Complete Guide to the British Dragonflies* (1986). I regard this work as pioneering as it is the first of its genre to include photographs of all the species, and particularly so in that the images were taken on film. With the onset of digital photography, almost all field observers can take useful shots, but McGeeney's task was far more difficult and required a much greater degree of fieldcraft and photographic expertise. Even so, the almost bewildering array of colour forms in the damselflies that I encountered on my forays into the field largely passed me by without name.

The last decade has seen the publication of several worthy successors, for example, Dave Smallshire and Andy Swash's *Britain's Dragonflies: A Field Guide to the Damselflies and Dragonflies of Great Britain and Ireland* (2004, 2010, 2014, 2018). In these particular works, there are excellent digital images of our British species, but also – and despite the title *Britain's Dragonflies* – many that have hitherto never been recorded from these shores. Although I thoroughly admire the first edition of this book and its two subsequent revisions, their beautiful production and the expertise of their authors, I have pondered if this represents too much information for the average British or Irish dragonfly enthusiast to digest.

The next era in the history of dragonfly identification guides came with the arrival of the artist Richard Lewington. His illustrations for the series of books *The Moths and Butterflies of Great Britain and Ireland* and others on the Lepidoptera are superb and indeed breath taking in their accuracy and texture. The dragonfly world has been fortunate indeed to have had Richard's artistic genius available for what have become the 'standard' field guides on their subject, *Field Guide to Dragonflies and Damselflies of Great Britain and Ireland* (Brooks & Lewington, 2004; Brooks, Cham & Lewington, 2014) and the 'sister' volume *Field Guide to the Dragonflies of Britain and Europe* (Dijkstra & Lewington, 2006). Both books contain excellent illustrations of the whole insect and its particular taxonomic features, as well as species accounts by several experts in their field. Moreover, the former deals only with those species already recorded from our islands and does not speculate on what may yet turn up in the future.

It would be remiss of me not to also mention *Guide to the Dragonflies and Damselflies of Ireland* (Thompson & Nelson, 2014). This wonderful book is a kind of 'hybrid' publication that uses the very best photographs of each species as

Introduction

well as Richard's artwork. For what it's worth, this is my personal favourite.

Interest in the Odonata has increased greatly in recent years. Indeed, many birdwatchers now spend the summer months in search of dragonflies and damselflies. This has generated a thirst for information regarding sites where the more localised species may be found. This was perhaps first addressed by Paul Hill and Colin Twist (1996) in their *Butterflies and Dragonflies: A Site Guide* (revised 1998). Their volume was the first to give extensive details of sites for each species, including O. S. Grid References, parking facilities and site maps. The information found within those pages was a revelation and allowed enthusiasts to find and enjoy seeing the scarcer species. The most recent similar guide was published in 2007 by Steve and Caroline Dudley and Andrew Mackay. *Watching British Dragonflies* contains nearly 200 pages of information on recommended sites to visit and, in this respect, may well never be surpassed.

Whilst there is clearly a wealth of literature currently available on where to go, what to look for, as well as to help us with identification of the British dragon- and damselflies, having worked and led tours for many competent naturalists over many years, I am frequently reminded that there is much scope for improving the 'user-friendliness' of such guides. Therefore, in light of this and with the primary aim to improve user-friendliness, but also with due deference to my learned colleagues and their previously published works on this group of insects, I therefore gladly submit to the reader this new, colour photographically illustrated identification guide to these unquestionably inspiring creatures.

How to use this book

The method devised here for identifying Odonata is similar in principle to that used in one of my previous works on the Lepidoptera: *British and Irish Pug Moths* (Riley & Prior, 2003). This entails dividing the insects into separate groups of superficially similar taxa regardless of their taxonomic status (i.e. species, form or sex). This results in the reader having only to refer to a small group containing, for example, red, blue or green damselflies rather than having to peruse all of the species in search of the specimen 'in hand'.

The procedure involves following the steps detailed below. In this example we shall use the finding of a bronze-green damselfly with a blue 'tail' at a reed-fringed pond in Norfolk during early July. It is resting with its wings half open.

1. Try to get photographs, or take comprehensive notes, of the insect. These should include details, from above and from the side, of the head, thorax, wingtips and the base and tip of the abdomen. If photographs are taken, do not try to fill the frame as this will narrow the depth of field. Also, try to remember to keep the plane of the subject parallel to that of the lens. This will help to keep the whole length of the insect in focus. A third tip is to approach with the sun behind one's back. This is particularly important when photographing the dragonflies as they will almost certainly be resting with their wings open towards the observer as they attempt to absorb the sun's warmth. This makes them easier to find, but care must always be taken to avoid casting a shadow over or close to the insect. To this end it is wise to remove one's hat so as to create as little shadow as possible. Photographs or a description of the habitat should also be taken.

2. Refer to the start of Chapter Two where there are representative images of each of the 13 groups of damselfly and decide which photograph the specimen in question most resembles. In this example it would be Group 3 '*Metallic bronze-green with a blue 'tail'*'.

3. Go to Group 3. At the beginning there are photographs of the two species dealt with therein: Emerald Damselfly and Scarce Emerald Damselfly.

4. The taxonomic features that separate the two species should then be read and compared with the accompanying close-up images of the various parts of the insects. For example, if the specimen in question is of comparatively robust build, has only half of the second abdominal segment blue, the pterostigmata are edged at each end with a fine, white line and the claspers at the tip of the abdomen are inwardly curved. These points would suggest that it is a Scarce Emerald Damselfly.

5. Refer to Chapter One and read the species account for that species. Check that the flight period, distribution, habitat preferences, detailed description

How to use this book

and behaviour provide a match. This should then confirm that the specimen is, indeed, a Scarce Emerald Damselfly.

The species accounts detailed in Chapter One are divided into the following parts:

Flight period
This gives the overall flight period as gleaned from the histograms provided by Cham *et al.* (2014). The optimum time for finding a species has been estimated from here and from my own field notes.

Distribution and status
An overview of national distribution is provided using as its basis the Watsonian Vice-counties of England, Wales and Scotland, and the Praeger Vice-counties of Ireland (see Appendix I). These boundaries have remained stable since their inception in 1852 and 1901, respectively, are likely to do so into the future and are not influenced by the vagaries of political decisions.

Habitat preferences
For the more specialised species a photograph of their typical habitat is provided as a field aid.

Description
There is a detailed description of the sexes and recognised forms for each species. Microscopic details, or those requiring the use of a powerful hand lens, are not discussed as this would usually mean handling the insect – a practice that I do not wish to encourage.

Similar species
Comparison with species of similar appearance allows a swift route to double-checking one's identifications.

Behaviour/field tips
Here you find observations on the behaviour of each species listed with tips on how to find them in the field. There are also precautionary notes where potentially dangerous sites are discussed.

Systematic Checklist
and Status of Species and described Forms recorded in Britain and Ireland

Category A: Resident

Willow Emerald Damselfly *Chalcolestes viridis* (Vander Linden, 1825)

Scarce Emerald Damselfly *Lestes dryas* Kirby, 1890

Emerald Damselfly *Lestes sponsa* (Hansemann, 1823)

Banded Demoiselle *Calopteryx splendens* (Harris, 1782)

Beautiful Demoiselle *Calopteryx virgo* (Linnaeus, 1758)

White-legged Damselfly *Platycnemis pennipes* (Pallas, 1771)
   Form *lactea* (Charpentier, 1825)

Small Red Damselfly *Ceriagrion tenellum* (Villers, 1789)
   Form *erythrogastrum* Sélys, 1876
   Form *intermedium* Sélys, 1876
   Form *melanogastrum* Sélys, 1876

Northern Damselfly *Coenagrion hastulatum* (Charpentier, 1825)

Irish Damselfly *Coenagrion lunulatum* (Charpentier, 1840)

Southern Damselfly *Coenagrion mercuriale* (Charpentier, 1840)

Azure Damselfly *Coenagrion puella* (Linnaeus, 1758)

Variable Damselfly *Coenagrion pulchellum* (Vander Linden, 1825)

Common Blue Damselfly *Enallagma cyathigerum* (Charpentier, 1840)

Red-eyed Damselfly *Erythromma najas* (Hansemann, 1823)

Small Red-eyed Damselfly *Erythromma viridulum* (Charpentier, 1840)

Blue-tailed Damselfly *Ischnura elegans* (Vander Linden, 1820)
   Form *violacea* Sélys, 1837*
   Form *infuscans* Campion & Campion, 1905
   Form *rufescens* Stephens, 1835
   Form *infuscans-obsoleta* Killington, 1924

Scarce Blue-tailed Damselfly *Ischnura pumilio* (Charpentier, 1825)
   Form *aurantiaca* Sélys, 1837

Large Red Damselfly *Pyrrhosoma nymphula* (Sulzer, 1776)
   Form *fulvipes* Stephens, 1835
   Form *melanotum* Sélys, 1876

Azure Hawker *Aeshna caerulea* (Ström, 1783)

Southern Hawker *Aeshna cyanea* (Müller, 1764)

Brown Hawker *Aeshna grandis* (Linnaeus, 1758)

Systematic Checklist

Common Hawker *Aeshna juncea* (Linnaeus, 1758)

Migrant Hawker *Aeshna mixta* Latreille, 1805

Norfolk Hawker *Anaciaeshna isosceles* (= *isoceles*) (Müller, 1767)

Emperor Dragonfly *Anax imperator* Leach, 1815

Hairy Dragonfly *Brachytron pratense* (Müller, 1764)

Club-tailed Dragonfly (= Common Clubtail) *Gomphus vulgatissimus* (Linnaeus, 1758)

Golden-ringed Dragonfly *Cordulegaster boltonii* (Donovan, 1807)

Downy Emerald Dragonfly *Cordulia aenea* (Linnaeus, 1758)

Northern Emerald Dragonfly *Somatochlora arctica* (Zetterstedt, 1840)

Brilliant Emerald Dragonfly *Somatochlora metallica* (Vander Linden, 1825)

White-faced Darter *Leucorrhinia dubia* (Vander Linden, 1825)

Broad-bodied Chaser *Libellula depressa* Linnaeus, 1758

Scarce Chaser *Libellula fulva* Müller, 1764

Four-spotted Chaser *Libellula quadrimaculata* Linnaeus, 1758
  Form *praenubila* Newman, 1833

Black-tailed Skimmer *Orthetrum cancellatum* (Linnaeus, 1758)

Keeled Skimmer *Orthetrum coerulescens* (Fabricius, 1798)

Black Darter *Sympetrum danae* (Sulzer, 1776)

Red-veined Darter *Sympetrum fonscolombii* (Sélys, 1840)

Common Darter *Sympetrum striolatum* (Charpentier, 1840
  Form *nigrescens* Lucas, 1912 'Highland Darter'

Ruddy Darter *Sympetrum sanguineum* (Müller, 1764)

* It appears likely that Sélys, in his 1837 publication, actually assigned the name *violacea* to a form of Common Blue Damselfly, but this has subsequently been used erroneously, under the authority of Sélys, as the title for the violet form of Blue-tailed Damselfly.

Category B: Annual Visitors and probable Colonists

Southern Emerald Damselfly *Lestes barbarus* (Fabricius, 1798)

Dainty Damselfly *Coenagrion scitulum* (Rambur, 1842)

Systematic Checklist

Southern Migrant Hawker *Aeshna affinis* Vander Linden, 1823

Lesser Emperor Dragonfly *Anax parthenope* (Sélys, 1839)

Vagrant Emperor Dragonfly *Anax* (= *Hemianax*) *ephippiger* (Burmeister, 1839)

Category C: Scarce or Assisted Visitors

Winter Damselfly *Sympecma fusca* (Vander Linden, 1820)

Marsh Bluetail *Ischnura senegalensis* (Rambur, 1842)

Green Darner *Anax junius* Drury, 1770

Yellow-legged Clubtail *Stylurus* (= *Gomphus*) *flavipes* (Charpentier, 1825)

Yellow-spotted Emerald *Somatochlora flavomaculata* (Vander Linden, 1825)

Scarlet Darter *Crocothemis erythraea* (Brullé, 1832)

Large White-faced Darter *Leucorrhinia pectoralis* (Charpentier, 1825)

Wandering Glider *Pantala flavescens* (Fabricius, 1798)

Yellow-winged Darter *Sympetrum flaveolum* (Linnaeus, 1758)

Banded Darter *Sympetrum pedemontanum* ([Müller, 1766])

Vagrant Darter *Sympetrum vulgatum* (Linnaeus, 1758)

Category D: Extinct Residents

Norfolk Damselfly *Coenagrion armatum* (Charpentier, 1840)

Orange-spotted Emerald Dragonfly *Oxygastra curtisi* (= *curtisii*) (Dale, 1834)

Anatomy

The following figures show the positions of the parts of the dragonfly and damselfly body that are referred to in the main substance of this text. These should be studied before reading the species descriptions.

Fig A: i. Postocular bar, **ii.** Postocular spot, **iii.** Antehumeral stripe, **iv.** Abdominal segment 2

Banded Demoiselle

CHAPTER ONE
Species Accounts of the resident Damselflies

Willow Emerald Damselfly *Chalcolestes viridis*	26
Scarce Emerald Damselfly *Lestes dryas*	28
Emerald Damselfly *Lestes sponsa*	30
Banded Demoiselle *Calopteryx splendens*	32
Beautiful Demoiselle *Calopteryx virgo*	34
White-legged Damselfly *Platycnemis pennipes*	36
Small Red Damselfly *Ceriagrion tenellum*	38
Northern Damselfly *Coenagrion hastulatum*	40
Irish Damselfly *Coenagrion lunulatum*	42
Southern Damselfly *Coenagrion mercuriale*	44
Azure Damselfly *Coenagrion puella*	46
Variable Damselfly *Coenagrion pulchellum*	48
Common Blue Damselfly *Enallagma cyathigerum*	50
Red-Eyed Damselfly *Erythromma najas*	52
Small Red-eyed Damselfly *Erythromma viridulum*	54
Blue-tailed Damselfly *Ischnura elegans*	56
Scarce Blue-tailed Damselfly *Ischnura pumilio*	58
Large Red Damselfly *Pyrrhosoma nymphula*	61

Willow Emerald Damselfly
Chalcolestes viridis (Vander Linden)

Fig. 1 Willow Emerald Damselfly. Titchwell NR, West Norfolk

Flight period
Usually from late June to late October. Optimum from early August and mid-September.

Distribution and status
A very recent colonist from Mainland Europe, this enigmatic species has subsequently established healthy populations in several English counties as far north as North-east Yorkshire and west as far as Oxfordshire. It has not yet been recorded from Scotland, Wales, Ireland, the Isle of Man or the Isles of Scilly.

Habitat preferences
Little specific detail is available regarding the species' requirements in the British Isles, but Askew (2004) states that in mainland Europe it breeds in ponds, lakes, canals and slow-flowing rivers where there are overhanging willows (*Salix* spp.). Eggs are laid in the branches of these and other trees and bushes. Fig 2 shows a typical English site for the species.

Description
Length c. 45mm.
Both sexes (male Fig 61; female Fig 62).
Head Metallic green with eyes brown above and pale greenish-brown below; lower facial markings blue (Fig 3).
Thorax From above, metallic green with one central and two fine, cream antehumeral stripes (Figs 67a & 67b). From the side, metallic green with two diagonal, yellow-green panels, the upper one of which is forked with upper 'tine' long and sometimes broken towards the tip (Figs 68a & b). Legs glossy black, paler below.

Fig. 2 Campsea Ashe, East Suffolk

Willow Emerald Damselfly

Fig. 3 Facial markings of Willow Emerald Damselfly

Abdomen From above, metallic green, turning bronze towards the tip, and with fine, pale intersegmental divisions (Figs 61 & 62). Segment 1 of the female cream with a pair of conjoined green spots, usually forming a continuous bar (Figs 67a & b). Superior male claspers inwardly curved and distinctively white (Fig 69a). From the side, metallic emerald green with lower half of segments 1 & 2 and half of segment 3 cream (Figs 68a & b).
Wings Clear with pterostigmata white or very pale brown (Fig 70a).

Similar species
Both sexes
See Damselfly Group 4.
Emerald Damselfly female and immature male (p 71, 73).
Scarce Emerald Damselfly female and immature male (p 71, 73).

Behaviour/field tips
Males hold vertical territories in tall waterside bushes such as Elder (*Sambucus nigra*) and willow (*Salix* spp.) saplings. Here they perch in exposed sunny positions and pursue other damselflies that stray too close. It appears not to become fully active until late morning. Before this time, it seems to be present only along the riverside and can then be difficult to locate and, if disturbed, often flies directly upwards into the canopy of nearby trees. Towards the end of the morning it may be found in profusion along hedgerows and the edges of woods adjacent to the water. Here, a light tap of the branches as one walks along will usually disturb the damselflies into a short flight. Once they have resettled, a cautious approach should allow good opportunities to photograph them.

Scarce Emerald Damselfly
Lestes dryas Kirby

Fig. 4 Scarce Emerald Damselfly Thompson Common, West Norfolk

Flight period
Usually from mid-June to late August. At sites where this species occurs alongside Emerald Damselfly, it usually begins to emerge a week or so earlier. Optimum throughout July.

Distribution and status
This is one of Britain's most localised species, though where it occurs it can be quite common.
England: Its two main strongholds are centred on the counties of West Kent, Surrey, South Essex and Middlesex in the south and West Norfolk in the north. Absent from the Isle of Man and the Isles of Scilly.
Ireland: Based mainly in the counties of Clare and South-east, West and North-east Galway. There are small populations elsewhere, and its superficial similarity to the Emerald Damselfly leads one to believe there are yet more to be discovered.
Wales and Scotland: Absent

Habitat preferences
The Scarce Emerald breeds in small, often quite deeply shaded, ponds and ditches that are surrounded by dense vegetation. Many of the pools at its inland sites are prone to extreme seasonal changes in water levels and are often dry during the summer. However, the species is adapted to cope with these situations and is also tolerant of the brackish conditions found at some of its coastal sites. Fig 5 shows a typical site at one of the Norfolk pingos.

Fig. 5 Thomson Common, West Norfolk

Scarce Emerald Damselfly

Description
Length *c*. 35mm.
Adult male (Fig 57)
Head Metallic green or bronze-green with eyes blue above and pale grey below (Fig 58b); lower facial markings pale brown.
Thorax From above, metallic green with a very fine (sometimes absent), pale central line (Fig 58b). From the side, segments 1 & 2 metallic green and segment 3 with pale-blue pruinescence. Legs glossy black, paler below.
Abdomen From above, metallic green with a pale-blue pruinescence on segment 1 and half of segment 2 at the base and half of segment 8 and all of segments 9 & 10 at the tip. The latter creates a conspicuous blue 'tail' (Fig 57). From the side this pattern is repeated. Inferior claspers broad and inwardly curved (Fig 59b).
Wings Clear with pterostigmata dark grey or black and edged at each end with white (Fig 60b).

Immature male (Fig 65)
As above, but abdomen with blue pruinescence replaced with metallic green; side of thorax with blue replaced with cream; eyes dark brown above and pale brown below (Figs 65; 67e; 68e).

Female (Fig 66)
Head As for immature male (Figs 66; 67e; 68e).
Thorax From above similar to male. From the side, segments 1 and upper half of segment 2 metallic green. Lower half of segment 2 with cream, forked marking with upper 'tine' short.
Abdomen From above, metallic green, turning bronze-green towards the tip and with two tiny, usually square, metallic spots on a cream background on segment 1 (Figs 66; 67f). These are often difficult to see in the field, and their shape can vary, so should not be used as a sole means of identification. From the side, segments 1 & 2 metallic bronze-green and lower half of segment 3 cream (Fig 68f).
Wings Similar to male.

Similar species
Adult male
See *Damselfly Group 3*.
Emerald Damselfly adult male (p **68**).

Adult female and immature male
See *Damselfly Group 4*.
Willow Emerald Damselfly (both sexes) (p **71**, **73**).
Emerald Damselfly female and immature male (p **73**).

Behaviour/field tips
Adults are rarely found far from their breeding ponds, though they do venture onto adjacent grassy meadows to feed. They are slow, weak fliers that spend much of their time perched on waterside vegetation with their wings held half open. They are easily disturbed into flight, but their green colouration can make them difficult to follow on the wing. As with the Emerald Damselfly, they are often easy to spot as they bask at the water's edge in the early morning sun and can sometimes be found together.

Emerald Damselfly
Lestes sponsa (Hansemann)

**Fig. 6 Emerald Damselfly
New Forest, South Hampshire**

Flight period
Usually late June to late September but may begin slightly earlier at southern localities. Optimum from mid-July to late August.

Distribution and status
Widespread and fairly common throughout most of the British Isles but absent from the Shetland Islands and the Isles of Scilly.

Habitat preferences
Restricted to standing water such as pools, ditches and canals with luxuriant marginal vegetation. Often found on heathland and moorland where there are acid bog pools. The species is also able to breed in brackish conditions. Figure 7 shows a windswept upland moorland pool on the Longmynd, Shropshire where the Emerald damselfly thrives.

Description
Length *c*. 35mm.
Adult male (Fig 56)
Head Metallic bronze-green with eyes blue above and pale grey below (Fig 58a); lower facial markings pale brown.
Thorax From above, metallic green with a very fine (if at all present), pale central and antehumeral lines (Fig 58a). From the side, segment 1 metallic green; segment 2 metallic green above and pale blue below; segment 3 pale blue. Legs glossy black, paler below.
Abdomen From above, metallic green, turning bronze-green towards the tip. Segments 1 & 2 and 9 & 10 with pale-blue pruinesence (Fig 59a). The latter

Fig. 7 The Longmynd, Shropshire

Emerald Damselfly

have a clear division between the blue and the metallic green of segment 8. From the side, the pattern of the dorsal surface is repeated. Inferior claspers fine and straight (Fig 59a).
Wings Clear with pterostigmata long and dark grey or black (Fig 60a).

Immature male (Fig 63)
As above, but with blue pruinesence replaced by metallic green; side of thorax with blue replaced with cream; eyes dark brown above and pale brown below (Fig 67c).

Female (Fig 64)
Head As for immature male.
Thorax From above, metallic green with fine, pale central and antehumeral stripes (Fig 67d). From the side, segments 1 and upper half of segment 2 metallic green with pale-cream antehumeral stripe clearly visible; lower half of segment 2 with cream, forked marking, the upper 'tine' of which is short; segment 3 cream; legs glossy black, paler below (Fig 68d).
Abdomen From above, metallic green with two tiny, usually tear-drop-shaped, metallic green spots on a cream background on segment 1 (Fig 67d). These are difficult to see in the field, and their shape can vary, so should not be used as a sole means of identification.
Wings As in male.

Similar species
Adult male
See *Damselfly Group 3*.
Scarce Emerald Damselfly adult male (p **68**).

Female and immature male
See *Damselfly Group 4*.
Willow Emerald Damselfly (both sexes) (p 71, **73**).
Scarce Emerald Damselfly female and immature male (p **73**).

Behaviour/field tips
This species and the very similar Scarce Emerald Damselfly (with which it sometimes flies) are rarely found far from their breeding areas, though they often hunt in adjacent grassland. They are usually the first damselflies to become active in the morning, and an early visit to the western side of a known breeding pond will often be rewarded with the sight of several of either, or both, perched on waterside plants taking advantage of the early morning sun. If disturbed into flight, their green colouration often makes them difficult to track. However, the blue 'tail' of the adult male is something of a giveaway as it can appear almost fluorescent against a dark background. All flights are low to the vegetation and short in duration.

Banded Demoiselle
Calopteryx splendens (Harris)

Fig. 8 Banded Demoiselle. Sculthorpe, West Norfolk

Flight period
Usually from late May to late August. Optimum from mid-June to early August.

Habitat preferences
Unpolluted, slow-moving stretches of streams, rivers or canals with thick bottom sediment. Has been recorded breeding in pools adjacent to such waters (Brooks, 2004; Dudley *et al.*, 2007). The area around these chosen sites is usually open with little shade. Adults hunt amongst low bankside vegetation, such as nettle and bramble beds, or along the edges of waterside meadows. Figure 9 shows a typical stronghold on the River Wensum in West Norfolk.

Distribution and status
England and Wales: Widespread and locally common south of a line approximately between South Lancashire in the west and North Northumberland in the east, including Anglesey, but absent from much of West Cornwall and South Devon. There are small outlying colonies in Cumberland. It has not been recorded from the Isles of Scilly.
Ireland: Widely distributed throughout.
Scotland and the Isle of Man: Absent.

Banded Demoiselle

Description
Length *c.* 45mm.
Male (Fig 50)
Head Metallic green with eyes dark red.
Thorax Metallic green. From above with fine, pale central and antehumeral stripes; legs black.
Abdomen Metallic blue, turning greenish-blue on segments 8–10. Inferior claspers broad and straight.
Wings Clear, each with a large, broad and diagnostic navy-blue patch.

Female (Fig 53)
Head and thorax As in the male but lacking fine central and antehumeral stripes.
Abdomen Metallic green with segments 1 & 2 and 8–10 metallic bronze (Fig 53), the latter with a faint, pale central stripe along the upper surface (Fig 55).
Wings Greenish tinged with white pseudo-pterostigmata (Fig 53).

Similar species
Male
None. The large, dark-blue wing patches on otherwise clear wings are unique to this species.

Female
See Damselfly Group 2.
Beautiful Demoiselle male and female (p **67**).

Behaviour/field tips
Very butterfly-like in flight. Flits gracefully and purposefully over the water, settling often on emergent vegetation in the middle of the stream or river and on bankside shrubs and bushes. Whilst so resting, they can be remarkably inconspicuous for such a large and boldly marked insect. Doubtless, the best way to see this species is simply to wait on the bank of a suitable site for a period of sunshine. If present, the damselflies will then suddenly appear as if from nowhere. When they return to their resting places, they are usually easy to approach and photograph. The females appear to fly most actively at the breeding sites during the late morning and early afternoon.

Fig. 9 Sculthorpe Mill, West Norfolk

Beautiful Demoiselle
Calopteryx virgo (Linn.)

Fig. 10 Beautiful Demoiselle. New Forest, South Hampshire

Flight period
Usually from mid-May to the end of August. Optimum from early June to late July.

Habitat preferences
Unpolluted, briskly flowing streams and rivers with a clean gravel or sand substrate. Stands of emergent, or bankside, vegetation are required for mating and resting. Often found in such places on moorland and heathland. It is shade tolerant and can often be seen flying beneath riverside trees. Figure 11 shows a typical site in the New Forest where this species can be seen in good numbers.

Distribution and status
England: Widespread and locally common south of a line approximately between Cheshire in the west and East Kent in the east, though it is absent from large areas of South Wiltshire and North and South Hampshire. Elsewhere in England there are colonies based in Westmorland and Cumberland in the west and North-east Yorkshire and Durham in the east. Absent from the Isle of Man and the Isles of Scilly.
Wales: Widespread throughout but localised in the island of Anglesey.
Scotland: Restricted approximately to the counties of West Inverness-shire and Argyll Main (including the islands of North, South and Mid Ebudes). Absent from the Western and Northern isles.
Ireland: Found south of a line approximately between North Kerry in the west and Dublin in the east with outlying populations centred in West Galway.

Description
Length *c.* 45mm.
Male (Fig 51)
Head Metallic green with eyes glossy dark brown.
Thorax Entirely (including the legs) metallic greenish-blue.
Abdomen Deep metallic blue, turning to bluish-green towards the tip. Inferior claspers fine and slightly inwardly curved.
Wings Almost entirely deep blue, appearing dark green or purple according to the angle of the light. The tips may fade with age and wear.

Female (Fig 52)
Head Similar to the male.
Thorax Similar to the male but with metallic-green legs.
Abdomen Metallic green, turning to bronze-green towards the tip. Segments

Beautiful Demoiselle

7–10 from above with a fine, but obvious, pale-brown central stripe (Fig 54).
Wings Tinged brown with small, white pseudo-pterostigmata (Fig 52).

Similar species
Male
None. The entirely dark, blue-tinged wings are unique to this species.

Female
See Damselfly Group 2.
Banded Demoiselle female (p **67**).

Behaviour/field tips
Very graceful and butterfly-like in flight. The species is very easy to find in the appropriate areas as the males fly actively and conspicuously over the water and its surface vegetation. However, the females are usually only seen at their breeding sites during the late morning and early afternoon. At other times they may be found in sheltered spots around trees and bushes away from the water's edge.

Fig. 11 Mill Lawn Brook, New Forest, South Hampshire

White-legged Damselfly
Platycnemis pennipes (Pallas)

Fig. 12 White-legged Damselfly. Atcham, Shropshire

Flight period
Late May to mid-August. Optimum period from mid-June to late July.

Distribution and status
England: Locally common south of a line approximately between Shropshire in the west and Cambridgeshire in the east. However, its distribution is rather patchy, and the insect appears to be absent from much of West and East Cornwall, South Somerset and Dorset in the west, South Hampshire and South Wiltshire in the south and East Anglia, South Essex and East Kent in the east. Absent from the Isle of Man and the Isles of Scilly.
Wales: Restricted mainly to the mid-eastern counties. It has not been recorded from Anglesey.
Scotland and Ireland: Absent.

Habitat preferences
This species requires slow-flowing sections of lowland rivers or canals with luxuriant emergent and bankside vegetation. It is usually intolerant of shade, preferring open, sunny situations or large, well-lit areas in woodlands. It has only rarely been recorded breeding at still water sites – a habitat that is utilized commonly on the continent (Brooks, 2004).

Description
Length *c.* 36mm.
Adult male (Fig 71)
Head Black with two fine, pale lines (blue in the adult) joining the eyes (Fig 71). Eyes pale brown, the upper half turning to blue in the adult (Fig 74a).
Thorax From above, black with fine, pale-blue central and antehumeral stripes. From the side, pale greenish-blue with black stripes along the segmental divisions. All legs with long spines on the fibia and tibia. Fibia broad, black on the upper surface and pale blue below. Tibia conspicuously broad and white or very pale blue with a fine, black central stripe (Fig 74a).
Abdomen Ground colour pale blue. From above, with a series of black, elongated, arrow-shaped central markings on segments 2–5; segment 6 with a pair of black posterior spots; segments 7–10 black with a fine, pale-blue central stripe.
Wings Clear with distinctive amber-coloured pterostigmata (Fig 74b).

Adult female (Fig 72)
Head and thorax Eyes blue above and pale brown below.
Abdomen Ground colour very pale green. From above, with a pair of long, fine, black lines, broadened at the posterior

White-legged Damselfly

end, on each of segments 2–9; segment 10 without markings. From the side, segments 2–8 each with a long, fine, black ventral line, broadened at the posterior end.
Wings Clear with pterostigmata amber.

Immature female (form *lactea*)
(Charpentier, 1825) (Fig 73)
Head and thorax Similar to adult.
Abdomen Pale milky white. From above, with a pair of small, black lateral dots on segments 2–6. Segments 7–10 black with a fine, milky-white central stripe. From the side, segments 2–6 each with a single black spot along both dorsal and lateral surfaces. Segments 7–10 black above and milky white below. This form was once thought to be a separate species (*Agrion lacteum* Charpentier).
Wings Clear with pterostigmata pale yellow.

Similar species
None. The very pale colour of the abdomen, broad, feather-like legs, amber pterostigma and pale-blue eyes combine to make this species unlike any other damselfly found in the British Isles.

Behaviour/field tips
At favoured sites, this is frequently the commonest species of damselfly on the wing. Large numbers may be encountered hunting and basking amongst dense bankside vegetation. In sunny conditions they are very active and have

Fig. 13 Ovipositing White-legged Damselfly. Ober Water in the New Forest, South Hampshire

a distinctive jinking flight pattern. The males are very easy to distinguish as they fly with their white, feather-like legs dangling conspicuously beneath them. When the sun is obscured by cloud, the insects quickly seek shelter in deep vegetation and can then be extremely difficult to find, or even put into flight. On the re-appearance of the sun, they quickly emerge as if from nowhere. In anything less than sunny conditions, one could very easily be forgiven for thinking the species was absent at a newly visited site.

Small Red Damselfly
Ceriagrion tenellum (Villers)

**Fig. 14 Small Red Damselfly.
New Forest, South Hampshire**

Flight period
Mid-June to mid-September. Optimum from late June to mid-August.

Distribution and status
This is one of Britain's most localised and rarest species of damselfly.
England: There are three main population centres. The first is based in Surrey and its surrounding counties, the second in South Hampshire and Dorset and the third in West and East Cornwall and North and South Devon. There is also a single tiny, outlying colony in West Norfolk. Absent from the Isle of Man and the Isles of Scilly.
Wales: Restricted to the counties west of a line approximately between Carmarthenshire in the south and Caernarvonshire in the north. It is also present in the island of Anglesey.
Scotland and Ireland: Absent.

Habitat preferences
The Small Red Damselfly is usually found in the acidic conditions of heathland and moorland where it breeds in shallow unshaded pools, seepages and very slow-moving streams with dense vegetation. It is also sometimes associated with similar situations in calcareous fens, mires and disused marl pits. It is so tolerant of shallow water that at some sites it is able to complete its life cycle in a 'pool' created by an animal's hoofprint.

Description
Length *c*. 30mm.
Male (Fig 77)
Head Bronze-black with frons and eyes red (Fig 79c).

**Fig. 15 Mill Lawn Brook.
New Forest, South Hampshire**

Small Red Damselfly

Fig. 16 Small Red Damselfly with prey. New Forest, South Hampshire

Thorax From above, bronze black with antehumeral stripes absent or reduced to very fine, white or yellowish line (Figs 79c & d). From the side, bronze-black with pale-yellow longitudinal stripes on segments 2 & 3 (Fig 80b). Legs red.
Abdomen Entirely red.
Wings Clear with pterostigmata red (Fig 81b).

Female
Polymorphic, with four forms: the nomenclature and 'type' descriptions are after Askew (2004).
- **Form *tenellum*** (Villers) [typical form]
Thorax As in the male but upperside with antehumeral stripes sometimes more obvious.
Abdomen As in the male but upperside of segment 1 with black spot at base, segments 2 & 3 with black spot at apex and base, segments 4–8 black and segments 9 & 10 very dark red.
Eyes, wings and legs As in the male.
- **Form *erythrogastrum*** Sélys
This is a homeochromic form in which the female colouring matches that of the male.
- **Form *intermedium*** Sélys
Similar to typical *tenellum* but with upperside of abdominal segments 2–5 (rather than 2–3) red.
- **Form *melanogastrum*** Sélys (Fig 116)
Upperside of abdomen entirely black.

Similar species
Adult male and female forms *tenellum*, *erythrogastrum* and *intermedium*
See Damselfly Group 6.
Large Red Damselfly (both sexes) (p 78, 79).

Female form *melanogastrum*
See Damselfly Group 11.
Large Red Damselfly f. *melanotum* (p 98, 99).

Behaviour/field tips
This fragile creature rarely strays far from its breeding quarters. It flies only in warm, sunny conditions and is intolerant of wind. It's extremely delicate build and dark colouration makes it very difficult to spot when flying and, despite being red in colour, is remarkably cryptic when settled. A great deal of patience is therefore needed to find this enigmatic species. Great care must also be taken not to damage, by trampling, its delicate habitat.

Northern Damselfly
Coenagrion hastulatum (Charpentier)

Flight period
Early May to early August. Optimum from early June to early July.

Distribution and status
Restricted to Scotland where there are population centres based in East Inverness-shire, South Aberdeenshire and Perthshire. Even in its strongholds it is rarely common.

Habitat preferences
Confined to shallow, sheltered pools surrounded by dense marginal vegetation such as sedges. These requirements are very specific and, if not maintained, the species quickly falls into decline. Conversely should they be created newly in the vicinity of known breeding sites, the Northern Damselfly is very adept at colonisation. Waters lacking fish are most suitable as the larvae are very prone to predation (Brooks, 2004).

Fig. 17 Northern Damselfly female. Abernethy Forest, East Inverness-shire

Description
Length *c.* 32mm.
Male (Fig 97)
Head From above, black with post-ocular spots narrow, triangular and pale blue. Between them is a pale-blue, post-ocular bar (Fig 101b). From the side, lower half of eyes bright green (Fig 102b).
Thorax From above, black with prominent blue antehumeral stripes (Fig 101b). From the side, blue with a fine, black stripe along segmental divisions 2 & 3 (Fig 102b). Legs black above and pale blue below (Fig 102b).
Abdomen From above, mainly blue with extensive black markings. At the base there is a black spot on segment 2 that usually resembles the spade in the suit of playing cards. (Figs 97; 101b). At the tip, segments 8 & 9 are blue. The latter has a pair of black dots which may be enlarged into bold spots (Fig 105b). From the side, there is an isolated horizontal, black bar on segment 2 (Fig 102b). The ventral surface is blue.
Wings Clear with pterostigmata black.

Female (Fig 126)
Head From above, black with post-ocular spots narrow, triangular and pale brownish green. Between them is a pale brownish-green, post-ocular bar (Fig 132a). From the side, lower half of eyes bright green (Fig 133a).

Northern Damselfly

Thorax From above, black with antehumeral stripes bold bluish green (Fig 132a). From the side, pea green with a fine, black stripe along segmental divisions 2 & 3 (Fig 133a). Legs black above and pale green below.
Abdomen From above, velvet black with fine, pale-green intersegmental divisions (Fig 17). From the side, pea-green with fine, dark intersegmental divisions (Fig 126).
Wings Clear with pterostigmata black (Fig 126).

Similar species
Male
See Damselfly Group 9.
Common Blue Damselfly male and female 'blue' form (p **86**, 93).
Azure Damselfly male and female 'blue' form (p **86**, 88).

Female
See Damselfly Group 13.
Azure Damselfly female 'green' form (p **104**, 106).
Blue-tailed Damselfly female form *infuscans* (p **104**, 108).

Behaviour/field tips
This species can be very difficult to spot in the field as both sexes are dark in colour and slight of build. They fly weakly amongst emergent vegetation and rarely venture over open water or above the height of the marginal plants. Great diligence is therefore needed to find them, and the use of a pair of binoculars is an advantage. The adults are most active between approximately 11am and 3pm. They are reluctant to fly in cloudy or windy weather.

Fig. 18 Abernethy Forest, East Inverness-shire

Irish Damselfly
Coenagrion lunulatum (Charpentier)

Flight period
Early May to late July. Optimum from late May to late June.

Distribution and status
Restricted to Ireland. In the north it is found in small, widely scattered colonies within a band centred on counties Leitrim, Fermanagh, Monaghan, Armagh and Down. In the Republic, the insect is much scarcer and restricted to very small colonies in the central and eastern counties. Brooks (2004) states that a quarter of all known colonies have disappeared since the species' discovery in Ireland in 1981. This decline seems to be attributable mainly to nutrient enrichment of the breeding sites and/or lowering of water levels. Fortunately, several of the colonies in Northern Ireland are on protected reserves that should escape such environmental mismanagement or negligence. The presence of predatory fish such as Rudd (*Scardinius erythrophthalmus* (Linn.)) and small Pike (*Esox lucius* Linn.) may also be detrimental to small isolated colonies.

Habitat preferences
Typically in small, shallow and sheltered mesotrophic pools and lakes with clear water and an abundance of floating and marginal vegetation. It also inhabits small bog-pools that have been formed after peat extraction.

Description
Length *c.* 32mm.
Male (Fig 98)
Head From above, black with post-ocular

**Fig. 19 Irish Damselfly.
Montiaghs Moss, County Antrim**

spots crescent shaped and bright blue (Fig 101c). The post-ocular bar is absent. From the side, lower half of eyes bright green (Fig 102c).
Thorax From above, black with prominent blue antehumeral stripes (Fig 101c). From the side, upper half blue with a fine, black stripe along segmental divisions 2 & 3. Lower half bright green (Fig 102c). Legs black above and pale blue-green below.
Abdomen From above, blue with extensive black markings that give a generally dark appearance in the field (Fig 98). At the base there is a black, forward-facing, crescent-shaped marking on segment 2 (Fig 101c). From the side, there is an isolated black bar on segment 2 (Fig 102c). The ventral surface is bright green.

Irish Damselfly

Fig. 20 Peat cutting at Montiaghs Moss, County Antrim

Wings Clear with pterostigmata black (Fig 19).

Female (Fig 77)
Head From above, black with post-ocular spots inconspicuous, crescent shaped and green (Fig 132b). Post-ocular bar absent. From the side, lower half of the eyes bright green (Fig 133b).
Thorax From above, black with antehumeral stripes greenish-brown (Fig 132b). From the side, upper half greenish-brown with a fine, black stripe along segmental divisions 2 & 3. Lower half bright green with dark intersegmental divisions (Fig 133b). The legs are black above and pale green below.
Abdomen From above mainly black with fine, green intersegmental lines (Fig 127). From the side, lower half bright green (Fig 133b).
Wings Clear with pterostigmata black (Fig 19).

Similar species
Male
See Damselfly Group 9.
Because of its extremely localised distribution, the only blue and black damselflies that live alongside this species are:
Azure Damselfly male and female 'blue' form (p **87**, **88**).
Variable Damselfly male and female 'blue' form (p **87**, **91**).

Female
See Damselfly Group 13.
Azure Damselfly female 'green' form (p **104**, **106**).
Blue-tailed Damselfly female form *infuscans* (p **104**, **108**).

Behaviour/field tips
Flights are usually short and discrete. The dark colouration and delicate build make this a difficult species to spot. However, the males habitually perch on floating vegetation and, with practice, can then be found fairly easily. The use of binoculars is recommended for identifying individuals at rest. The females spend most of their time away from the water amongst dense, low vegetation. The adults are most active an hour or so either side of midday. During dull weather they hide deep in waterside vegetation and are reluctant to fly, even when disturbed.

Southern Damselfly
Coenagrion mercuriale (Charpentier)

Fig. 21 Southern Damselfly. New Forest South Hampshire

Flight period
Mid-May to early August. Optimum from early June to mid-July.

Distribution and status
This is one of Britain's rarest and most localised damselflies. It is protected under the British Wildlife and Countryside Act, 1981, and must not be caught or handled without a permit.
England: Restricted mainly to South Devon, South Hampshire and the south-easternmost parts of Dorset. Absent from the Isle of Man and the Isles of Scilly.
Wales: Confined mainly to the counties of Pembrokeshire, Carmarthenshire and the westernmost parts of Glamorgan. It is also present in small numbers in the island of Anglesey.
Scotland and Ireland: Absent.

Habitat preferences
Usually found on gravel-bottomed streams that traverse heathland, though those running over calcareous soil are used occasionally. It is intolerant of shade, preferring more open, warm situations. Such conditions are often created and maintained by the use of grazing stock.

Description
Length *c.* 30mm.
Male (Fig 96)
Head From above, black with post-ocular spots broad, triangular and pale blue. Post-ocular bar pale blue (Fig 101a). From the side, the lower half of the eyes is pale blue (Fig 102a).
Thorax From above, black with antehumeral stripes bold and blue (Fig 101a). From the side, blue with a fine, black stripe along segmental divisions 2 & 3 (Fig 102a). The legs are black above and pale blue below (Fig 102a).
Abdomen From above, blue with black markings (Fig 96). At the base there is a black mark on segment 2 that usually resembles a mushroom beneath an upward-pointing crescent, the whole being similar to the symbol for the planet Mercury (Fig 101a). At the base, a black, rectangular shape with two fine, forward-pointing spikes. From the side, blue with the edges of the dorsal markings still visible (Fig 21).
Wings Clear with pterostigmata short and black, sometimes with paler edges.

Female (Fig 128)
Head Black with post-ocular spots broad,

Southern Damselfly

triangular and pale green. Post-ocular bar pale green (Fig 132c).
Thorax From above, black with antehumeral stripes pale green (Fig 132c). From the side, pale green with a fine, black stripe along segmental divisions 2 & 3 (Fig 133c). Legs black above and pale green below.
Abdomen From above, black with pale-green intersegmental divisions (Fig 128). From the side, pale green (Fig 21).
Wings Clear with pterostigmata dark grey or black.

Similar species
Male
See Damselfly Group 9.
Azure Damselfly male and female 'blue' form (p **85**, **88**).
Common Blue Damselfly male and female 'blue' form (p **85**, **93**).

Female
See Damselfly Group 13.
Azure Damselfly female 'green' form (p **106**).
Scarce Blue-tailed Damselfly female (p **106**, **107**).
Blue-tailed Damselfly female f. *infuscans* (p **106**, **108**).

Behaviour/field tips
This is very much a sun-loving species that should be sought along the edges of streams through areas where there is little shade. It can often be found feeding around brambles and hawthorns in scrubby areas a little distance from water and it occasionally wanders some way into surrounding heathland. It is most active during the late morning and early afternoon but can often be disturbed from vegetation earlier in the day if the weather is warm.

Fig. 22 Mill Lawn Brook, South Hampshire

Azure Damselfly
Coenagrion puella (Linn.)

Fig. 23 Azure Damselfly. Upton Fen, East Norfolk

Flight period
Usually from mid-May to mid-August. Optimum from late May to late July.

Distribution and status
Widespread and very common throughout much of England, Wales, Ireland and the southern half of Scotland. Absent from the Isle of Man and the Isles of Scilly.

Habitat preferences
The Azure Damselfly can be found in a wide variety of habitats but prefers smaller, sheltered ponds and ditches where there is an abundance of emergent aquatic vegetation.

Description
Length *c.* 32mm.

Male (Fig 99)
Head From above, black with post-ocular spots triangular and blue (Fig 101d). Post-ocular bar absent. From the side, eyes black above with lower half blue (Fig 102d).
Thorax From above, black with ante-humeral stripes blue (Fig 101d). From the side, blue with a fine, black stripe along segmental divisions 2 & 3 (Fig 102d). Legs black above and pale blue below.
Abdomen From above, blue (paler with a lilac tint in immature) with black markings; segment 7 entirely black (Fig 99). At the base, segment 2 with a small, black marking resembling a fine 'U' or drinking tumbler (Fig 101d). At the tip, segment 9 with a bold, black marking resembling a crown (Fig 105d).
Wings Clear with pterostigmata dark grey or black.

Female 'blue' form (Figs 100; 111)
Head From above, similar to the male but blue markings paler (Figs 108a; 113b). From the side, lower half of eyes pale bluish-green (Figs 109a; 114b).
Thorax From above, similar to the male but blue markings slightly paler (Figs 108a; 113b). From the side, markings as in the male but ground colour pale green or blue (Figs 109a; 114b). Legs black above and pale bluish-green below.
Abdomen From above, blue with black markings far more extensive than those of the male. Sometimes so much so that only blue intersegmental divisions remain (Figs 100; 111). At the base, segment 2 with thistle-head-shaped

Azure Damselfly

marking. From the side, lower half pale bluish-green.
Wings As in the male.

Female 'green' form (Fig 129)
Head From above, similar to the 'blue' form but with post-ocular spots green (Fig 132d). From the side, ground colour green (Fig 133d).
Thorax Similar to 'blue' form but antehumeral stripes green (Fig 132d). Legs similar to 'blue' form (Fig 133d).
Abdomen From above, black with fine, pale-green intersegmental divisions (Fig 129). From the side, lower half ground colour pale green (Fig 23).
Wings Similar to male and female 'blue' form.

Similar species
Male and female 'blue' form
See Damselfly Group 9.
Southern Damselfly male (p 85, **88**).
Northern Damselfly male (p 86, **88**).
Irish Damselfly male (p 87, **88**).
Common Blue Damselfly male and female 'blue' form (p **88**, 93).
Variable Damselfly male and female 'blue' form (p **88**, 91).

Female 'blue' form
See Damselfly Group 10.
Small Red-eyed Damselfly female/immature male (p **97**).

Female 'green' form
See Damselfly Group 13.
Northern Damselfly female (p 104, **106**).
Irish Damselfly female (p 104, **106**).
Southern Damselfly female (p **106**).
Scarce Blue-tailed Damselfly female (p **106**, 107).
Blue-tailed Damselfly f. *infuscans* (p **106**, 108).
Red-eyed Damselfly female/immature male (p **106**, 111).

Behaviour/field tips
Where it occurs, this species is often very common and should pose no problems to find. Unlike the Common Blue Damselfly, and perhaps because of its more delicate build, it tends to avoid flying over open water – preferring the shelter of waterside vegetation. It is often found along woodland tracks some distance from the breeding grounds.

Fig. 24 Sparham Pools, East Norfolk

Variable Damselfly
Coenagrion pulchellum (Vander Linden)

Fig. 25 Variable Damselfly. Wicken Fen, Cambridgeshire

Flight Period
Usually mid-May to early August. Optimum from late May to late July.

Distribution and status
The Variable Damselfly has a very patchy distribution in the British Isles and, at most of its sites, it is often uncommon.
England: Absent from the extreme south-west but otherwise occurs in favoured areas south-east of a line approximately between North Somerset in the west and North Lincolnshire in the east. There are further small outlying populations centred around Cheshire, Shropshire, Cumberland and Co. Durham. It has not been recorded from the Isle of Man or the Isles of Scilly.
Scotland: Found mainly in the south-western counties of Dumfriesshire, Kirkcudbrightshire and Wigtownshire with a few further reports from Argyll Main.
Wales: Occurs mainly in those counties adjoining the south coast and also in the island of Anglesey.
Ireland: Widespread and common throughout, though more localised in the south-eastern counties.

Habitat preferences
Usually standing water such as ditches, ponds or canals. Occasionally in very slow-flowing rivers. As it is a weak flyer, it requires the shelter afforded by tall marginal vegetation such as reeds and adjacent bushes or hedges.

Description
Length *c.* 32mm.
Male (Fig 103)
Head From above, black with post-ocular spots crescent shaped and blue (Fig 101e) or bronze-brown in immatures (Fig 108c). Post-ocular bar blue (Fig 101 e). From the side, eyes dark above with lower half blue (Fig 25).

Fig. 26 Strumpshaw Fen, East Norfolk

Variable Damselfly

Thorax From above, black with blue antehumeral stripes (bronze-brown in immatures (Fig 108c)) incomplete and forming a pair of exclamation marks (Fig 101e; Fig 102e). From the side, blue (pale blue and lilac tinted in immatures) with a fine, black stripe along segmental divisions 2 & 3 (Fig 102e). Legs black above and pale blue below.
Abdomen From above, blue with black markings. Base with black shape resembling a stemmed wine goblet on segment 2 (Fig 101e). Tip with a black, rectangular spot with three forward-pointing spikes (Fig 105e). From the side, blue above and black below (Fig 25).
Wings Clear with pterostigmata dark grey or black.

Female 'blue' form (Fig 104)
Head As in the male but blue markings usually paler (Fig 108b).
Thorax As in male but blue markings much paler. Antehumeral stripes complete and pale blue (Fig 108b).
Abdomen From above, blue but with black markings much more extensive than in male. Base with thistle-head-shaped marking on segment 2 (Fig 108b). From the side, similar to male but with black markings more extensive.
Wings As in the male.

Female 'dark' form (Fig 110)
Head and thorax As in the female 'blue' form (Fig 113a).
Abdomen From above, almost entirely black with fine-blue intersegmental divisions that become more pronounced towards the tip (Fig 110). From the side, black with blue ventral surface visible.
Wings As in male and female 'blue' form.

Similar species
Its geographical range does not overlap that of Northern Damselfly.
Male and female 'blue' form
See Damselfly Group 9.
Southern Damselfly male (p 85, **91**).
Irish Damselfly male (p 87, **91**).
Azure Damselfly male and female 'blue' form (p 88, **91**).
Common Blue Damselfly male and female 'blue' form (p **91**, 93).

Female 'dark' form
See Damselfly Group 10.
Azure Damselfly female 'blue' form (p **96**, 97).
Small Red-eyed Damselfly female (p **96**, 97).

Behaviour/field tips
Its superficial resemblance to several other damselflies and its rather secretive behaviour, can make this a difficult species to find. It tends to avoid even the slightest breeze and so should always be sought amongst the shelter of plants growing along the water's edge or around nearby vegetation. Its slender build, combined with more extensive black markings, make the insect appear darker than its 'relatives', but this should only be used as a general guide. There is simply no substitute for examining closely all of the blue and black, or dark-coloured damselflies encountered.

Common Blue Damselfly
Enallagma cyathigerum (Charpentier)

Fig. 27 Common Blue Damselfly. New Forest, South Hampshire

Flight period
Usually between late May and early September. Optimum from early June to mid-August.

Distribution and status
A very common species throughout the British Isles, including the Outer Hebrides and the Shetland Islands. In the latter, it is the only resident species of damselfly.

Habitat preferences
Can be found in a very wide variety of open, still and slow-flowing aquatic habitats.

Description
Length *c*. 35mm.
Male (Fig 106; Fig 27)
Head From above, black with post-ocular spots crescent shaped and blue. The post-ocular bar is pale blue but sometimes inconspicuous (Fig 101f). From the side, lower half of eyes blue (Fig 102f).
Thorax From above, black with antehumeral stripes broad and blue (Fig 101f). From the side, blue with a single short, black line along segmental division 3 (Fig 102f). Legs black above and pale blue below (Fig 27).
Abdomen From above, blue with black markings. At the base, a black mark on segment 2 that resembles the pawn in a chess set (Fig 101f). At the tip, segments 8 & 9 entirely bright blue (Fig 105f). From the side, blue with black inter-segmental divisions (Fig 27).
Wings Clear with pterostigmata dark grey or black.

Female 'blue' form (Fig 107)
Head As in the male.
Thorax As in the male.

Common Blue Damselfly

Abdomen From above, blue with extensive black markings. At the base, segment 2 with black, thistle-head-shaped mark. Segments 3–7 each with a large, black marking that resembles an upward-pointing bomb (Fig 107; 108d). From the side, blue with edges of dorsal markings visible.
Wings As in the male.

Female 'brown' form (Fig 120)
Head From above, black with eyes, post-ocular spots and post-ocular bar either orange or greenish-brown (Figs 120; 122a). From the side, eyes brown or greenish-brown.
Thorax From above, black with antehumeral stripes either orange or greenish-brown. From the side, pale brown with a short, black stripe along segmental division 3 (Fig 123a). Legs black above and pale brown below (Fig 123a).
Abdomen From above, pale brown or greenish-brown with extensive black markings. At the base, segment 2 with black, thistle-head-shaped mark. Segments 3–7 each with a large, black marking that resembles an upward-pointing bomb (Fig 120). From the side, pale brown or greenish-brown with edges of dorsal markings visible.
Wings Clear with pterostigmata varying from pale grey to black.

Similar species
Male and female 'blue' form
See *Damselfly Group 9*.
Southern Damselfly male (p 85, **93**).
Northern Damselfly male (p 86, **93**).

Fig. 28 Emerging Common Blue Damselfly, South Hampshire

Irish Damselfly male (p **93**).
Azure Damselfly male and female 'blue' form (p 88, **93**).
Variable Damselfly male and female 'blue' form (p 91, **93**).

Female 'brown' form
See *Damselfly Group 12*.
Scarce Blue-tailed Damselfly female f. *aurantiaca* (p **101**).
Blue-tailed Damselfly female f. *infuscans-obsoleta* (p **101**, 103).

Behaviour/field tips
As its name suggests, the Common Blue Damselfly is indeed a very common species. At some lakes it can often be seen in hundreds flying low and strongly over the surface of the water. They sometimes are so abundant that they appear as a blue 'mist'. In the early part of the day, dozens can often be disturbed from vegetation adjacent to the breeding site. This species should pose few problems to find.

Red-eyed Damselfly
Erythromma najas (Hansemann)

Fig. 29 Red-eyed Damselfly. New Forest, South Hampshire

Flight period
Usually late May to early August. Optimum from early June to late July.

Distribution and status
England: Locally common south of a line approximately between Cheshire in the west and South-east Yorkshire in the east. However, apart from in South Devon, it is absent from much of the south-west. It is absent from the Isle of Man and the Isles of Scilly.
Wales: Restricted to the counties of the east and north-east. It has not been recorded in the island of Anglesey.
Scotland and Ireland: Absent.

Habitat preferences
Ponds, lakes, canals and very slow-moving rivers where there is an abundance of floating vegetation, such as the leaves of water lilies or mats of algae.

Description
Length *c.* 35mm.
Adult male (Fig 82)
Head Black with eyes dark red (Figs 84a; 87a).
Thorax From above, entirely black and lacking antehumeral stripes (Fig 84a). From the side, mainly blue with a black longitudinal stripe along segmental divisions 2 & 3. Legs dark grey or black (Fig 87a).
Abdomen From above, black with segments 1, 9 & 10 bright blue (Fig 85a). From the side, black with ventral surface dark green. Segments 1, 9 & 10

Red-eyed Damselfly

bright blue (Fig 87a).
Wings Clear with pterostigmata brownish-grey, edged with black (Fig 86a).

Immature male
Very similar to female (Figs 134; 132g; 133g) but lacking antehumeral stripes.

Female (Fig 134).
Head Mainly black with eyes dark brown (Fig 132g). From the side, eyes dark brown above and greenish-brown below (Fig 133g).
Thorax From above, black with pterostigmata short, fine and pale green (Fig 132g). From the side, green with a black line along segmental divisions 2 & 3. Legs black (Fig 133g).

Fig. 30 Lily pond at Pensthorpe West Norfolk

Abdomen From above, black with two blue segmental divisions near the tip (Fig 134). From the side, black above and green below (Fig 134).
Wings Clear with pterostigmata grey edged with black.

Similar species
This species' absence from Ireland and Scotland precludes confusion with Irish Damselfly and Northern Damselfly, respectively. Further, its habitat references make it very unlikely to be found alongside Scarce Blue-tailed Damselfly and Southern Damselfly.

Male
See Damselfly Group 7.
Small Red-eyed Damselfly male (p **80**).

Female and immature male
See Damselfly Group 13.
Azure Damselfly female 'green' form (p 106, **111**).
Blue-tailed Damselfly f. *infuscans* (p 108, **111**).

Behaviour/field tips
This species has a fast and powerful flight compared with most other damselflies and can often be seen hovering. However, it spends much of its time perched on floating leaves or algae and then offers good opportunities for study and photography. The best approach is simply to scan such places using binoculars or a birdwatching telescope. Immature individuals can sometimes be found in abundance amongst waterside bushes such as brambles and hawthorns.

Small Red-eyed Damselfly
Erythromma viridulum (Charpentier)

Fig. 31 Small Red-eyed Damselfly male. Sculthorpe, West Norfolk

Flight period
Early July to mid-September. Optimum from late July to late August.

Distribution and status
England: Since its discovery in Britain in 1999, this species has now become well established and locally common with records extending south-east of a line between West Cornwall in the west and South Northumberland in the east. There are also records from Cheshire and South Lancashire. New sites are being discovered annually and the resident population is supplemented some years by immigrants from Continental Europe. It has not yet been recorded in the Isle of Man or the Isles of Scilly.
Wales: It has been reported in recent years in Carmarthen, Glamorgan and Monmouthshire.
Scotland and Ireland: Absent.

Habitat preferences
Still, eutrophic waters such as sheltered ponds, ditches and canals with an abundance of floating vegetation, such as Hornwort (*Ceratophyllum demersum*) or mats of algae. Sometimes found in shallow, sheltered areas of larger water bodies where there is a thick sediment layer of decaying leaves amongst which the larvae can thrive.

Description
Length *c*. 30mm.
Male (Fig 31)
Head From above, black with a narrow, bright blue 'collar' and bright, tomato-red eyes (Fig 84b). From the side, eyes startlingly bright tomato red (Fig 87b).
Thorax From above, dark brown or black with antehumeral stripes yellow and tapered to a point towards the abdomen (Fig 84b). These are sometimes broken to form an exclamation mark. From the side, blue with a black stripe along segmental divisions 2 & 3. The former is often broken to leave an isolated black spot (Fig 87b). Legs black above and pale blue beneath.

Small Red-eyed Damselfly

Fig. 32 Small Red-eyed Damselfly female. Sculthorpe, West Norfolk

Abdomen From above, mainly black. At the base, segment 1 blue (Fig 84b). At the tip, sides of segment 8 and all of segment 9 blue. Segment 10 blue with a distinctive black cross (Fig 85b). From the side, mainly black. At the base, segments 1, 2 and the lower half of segment 3 blue (Fig 87b). At the tip, lower half of segment 8 and the whole of segments 9 & 10 almost entirely blue (Fig 86b).
Wings Clear with pterostigmata pale brownish-grey with darker edges (Fig 86b).

Female (Fig 112; Fig 32)
Head From above, black with a fine, bright blue 'collar' and brown eyes (Fig 113c). From the side, eyes brown above and green below (Fig 114c).
Thorax From above, dark brown or black with antehumeral stripes greenish-yellow and unbroken (Fig 113c). From the side, similar to the male (Fig 114c).
Abdomen From above, black with pale-blue segmental divisions (Fig 112). From the side, black with segments 1, 2 and half of segment 3 blue (Fig 32).
Wings Clear with pterostigmata pale grey with darker edges (Fig 32).

Similar species
Male
See Damselfly Group 7.
Red-eyed Damselfly male (p **80**).

Female
See Damselfly Group 10.
Variable Damselfly female 'dark' form (p **96**, **97**).
Azure Damselfly female 'blue' form (p **97**).

Behaviour/field tips
The males of this species spend a great deal of time resting on floating vegetation and can then be found quite easily by scanning with binoculars or a field telescope. The females are more often seen in rough vegetation near the breeding site and only visit the water to mate and lay eggs.

Blue-tailed Damselfly
Ischnura elegans (Vander Linden)

Fig. 33 Blue-tailed Damselfly. Reedham, East Norfolk

Flight period
From early May to mid-September. Optimum from early June to late August.

Distribution and status
Found commonly throughout most of the British Isles except for parts of central and northern Scotland and the Shetland Islands.

Habitat preferences
This very adaptable species is able to utilise most kinds of standing or slow-flowing waters, including brackish, and is even tolerant of low levels of pollution. It is often one of the first damselflies to colonise newly created ponds.

Description
Length *c*. 30mm.
Adult male (Fig 33; Fig 88)
Head From above, mainly black with post-ocular spots rounded triangular and blue (Fig 33). From the side, eyes black above with lower half blue (Fig 88).
Thorax From above, black with ante-humeral stripes bold and blue (Fig 33; Fig 88). From the side, blue with a short, black stripe along segments 2 & 3 (Fig 88). Legs black above and pale blue below.
Abdomen From above, mainly black. At the tip, segment 8 entirely blue (Fig 33). From the side, mainly black above and greenish-yellow beneath (Fig 95a). At the base, segments 1& 2 black above and blue below (Fig 88). At the tip, segments 7 & 9 black above and blue below, and segment 8 entirely blue (Fig 88).
Wings Clear with pterostigmata long, rhomboidal and conspicuously bi-coloured black and very pale brown, grey or white (Fig 33; Fig 95a).

Immature male (Fig 89)
Similar to adult but with blue markings on head and thorax replaced with green.

Female
There are five discrete colour forms of the female (originally described by Killington, 1924) each of which is identifiable in the field. They might be expected to be found throughout the species' range.
• **Typical form**
Very similar to the adult male.
• **Form *violacea*** Sélys (Fig 90)
Similar to typical form but with blue on

Blue-tailed Damselfly

head, thorax and base of abdomen replaced with violet.
- **Form *infuscans*** Campion & Campion (Fig 131)
Head, thorax and base of abdomen similar to typical form but with blue replaced with yellowish-brown. Tip of abdomen with blue on segment 8 replaced by brown (Fig 136).
- **Form *rufescens*** Stephens (Fig 91)
Similar to typical form but with blue on head, thorax and base of abdomen replaced with pink. Thorax with dark central stripe only.
- **Form *infuscans-obsoleta*** Killington (Fig 124)
Similar to f. *rufescens* but with pink on head, thorax and base of abdomen replaced with orange-brown (Figs 122c; 123c).

Similar species

Male and typical form of female
See *Damselfly Group 8*.
Scarce Blue-tailed Damselfly male (p 82, **84**).

Female form *violacea* and form *rufescens*
Unlike any other species.

Female form *infuscans*
See *Damselfly Group 13*.
Northern Damselfly female (p 104, **108**).
Irish Damselfly female (p 104, **108**).
Southern Damselfly female (p 106, **108**).
Azure Damselfly female 'green' form (p 106, **108**).
Scarce Blue-tailed Damselfly adult female (p 107, **108**).
Red-eyed Damselfly female/immature male (p **108**, 111).

Female form *infuscans-obsoleta*
See *Damselfly Group 12*.
Common Blue Damselfly female 'brown' form (p 101, **103**).
Scarce Blue-tailed Damselfly female f. *aurantiaca* (p 101, **103**).

Behaviour/field tips

With its dark colouration, delicate build and habit of flying low amongst bankside vegetation, this can sometimes be an awkward species to spot. However, it is usually common enough at its chosen locations to make success likely with patient searching. One should, perhaps, concentrate on watching for the conspicuous blue 'tail-light' moving against a dark background as it has an almost fluorescent quality. Although it spends most of its time close to its breeding grounds, individuals looking for new areas to colonize can often be found considerable distances from water.
The Blue-tailed Damselfly is very tolerant of cool, dull conditions and, provided it can find areas of shelter, will even remain active during windy weather. It is also one of the first species of damselfly to stir from its roosting sites in the early morning. It is not uncommon to see it flying at 7am or earlier.

Scarce Blue-tailed Damselfly
Ischnura pumilio (Charpentier)

Fig. 34 Scarce Blue-tailed Damselfly. New Forest, South Hampshire

Flight period
From late May until the end of August. Optimum from mid-May to late July.

Distribution and status
A very localised and generally uncommon species. Even at its favoured localities numbers are usually low.
England: Its two strongholds are in South Hampshire and Dorset in the south, and West and East Cornwall, South Devon and the extreme south-west of North Devon in the south-west. There are small colonies elsewhere in southern central England and West Norfolk, usually at man-made sites such as quarries. It is absent from the Isle of Man and the Isles of Scilly.
Scotland: Absent.
Wales: Widespread south-west of a line approximately between Caernarvonshire in the north and Monmouthshire in the south. It is also present in the island of Anglesey.
Ireland: Restricted to the counties of the north-west and south-east with outlying colonies in Clare.

Habitat preferences
At natural sites, this species occurs in valley mires and flushes where the water is shallow and mainly clear of vegetation. In such places regular animal activity helps to maintain these conditions. At man-made sites, such as mineral extraction workings, it is able to utilize areas where spring-fed seepages occur in tandem with vehicular disturbance that produces shallow pools or water-filled tracks. Such sites are usually transient. Because of these very specific requirements, the species is prone to large fluctuations in population size and frequent local extinctions. However, it appears to have a (apparently random) migratory strategy that sometimes enables it to establish new colonies.

Description
Length *c*. 30mm.
Adult male (Fig 92)
Head From above, mainly black with post-ocular spots rounded triangular and blue. From the side, eyes black above with lower half blue (Fig 92).
Thorax From above, mainly black with antehumeral stripes blue. From the side,

Scarce Blue-tailed Damselfly

mainly blue with a short, black stripe along segmental divisions 2 & 3. Legs black above and pale blue below (Fig 92).
Abdomen From above, mainly black. At the tip, two thirds of segment 8 black with the part adjacent to segment 9 blue. Segment 9 blue with a horizontal pair of black dots or spots (Fig 95b). From the side, mainly black above and greenish-yellow or bluish-yellow below. At the base, segments 1 & 2 black above and blue below (Fig 92). At the tip, lower half of segments 7 & 8 blue below and segment 9 entirely blue (Fig 95b).
Wings Clear with pterostigmata short, rhomboidal and clearly bicoloured black and very pale brown, grey or white (Fig 95b).

Immature male (Fig 93)
As in mature male, but blue on head, thorax and basal segments of abdomen replaced with pale green.

Female
As well as the typical form, there is an immature colour phase (f. *aurantiaca*) that is very distinctive in appearance and is described below. Also, there is a very scarce andromorhic form of the female that closely resembles the male in colouration (Fig 35).
• **Adult female** (Fig 130)
Head From above, anterior half of eyes black. Posterior half of eyes and ocular spots pale green (Fig 132e). From the side, eyes mainly black above and pale green below (Fig 34; 133e).
Thorax From above, mainly green with a broad, dark-brown central stripe and a fine, black line along segmental division 1 (Fig 132e). From the side, mainly green with a short, fine, black line along the bases of segmental divisions 2 & 3 (Fig 133e). Legs black above and pale green below (Fig 133e).
Abdomen From above, mainly black with pale-green intersegmental divisions (Fig 130). From the side, black above and pale green below. Legs black above and pale green below (Fig 133e).
Wings Clear with pterostigmata short, rhomboidal and bicoloured dark brown and very pale brown (Fig 135a).
• **Form *aurantiaca*** Sélys, 1837 (Fig 121)
• **Female andromorphic form** (Fig 35)

Fig. 35 The very rare female andromorphic form of Scarce Blue-tailed Damselfly. Winterton, East Norfolk

Scarce Blue-tailed Damselfly

Head and thorax Markings similar to adult female but ground colour bright orange (Fig 122b). Legs bright orange (Fig 123b).
Abdomen From above, black but with segments 1, 2 and upper part of 3 bright orange (Fig 121). From the side, mainly black above and bright orange below (Fig 125b).
Wings Clear with pterostigmata pale orange brown and not conspicuously bicoloured (Fig 125b).

Similar species
Adult male
See *Damselfly Group 8*.
Blue-tailed Damselfly adult male (p **82**, **84**).

Adult female
The geographical distributions of Northern Damselfly (Scotland only) and Irish Damselfly (Ireland only) preclude confusion with this species.
See *Damselfly Group 13*.
Southern Damselfly female (p 106, **107**).
Azure Damselfly female 'green' form (p 106, **107**).
Blue-tailed Damselfly f. *infuscans* (p **107**, 108).
Red-eyed Damselfly female & immature male (p **107**).

Female form *aurantiaca*
See *Damselfly Group 12*.
Common Blue Damselfly female 'brown' form (p **101**).
Blue-tailed Damselfly female f. *infuscans-obsoleta* (p **103**).

Behaviour/field tips
This species is reluctant to fly during cool, cloudy or windy conditions. Even in good weather it can be difficult to find as it tends to fly and rest amongst the shelter of low-growing vegetation. However, the males actively pursue damselflies of any species that venture too close to their chosen perch and, when doing so, can be spotted fairly easily. It is then a matter of distinguishing the individual in question from the much more common Blue-tailed Damselfly. Females and immatures tend to be more secretive and spend most of their time away from water whilst foraging in nearby scrubby areas. It therefore follows that this is where to search for the spectacular immature form *aurantiaca*.

Large Red Damselfly
Pyrrhosoma nymphula (Sulzer)

Flight period
Usually early May to mid-July in the south, though it is sometimes seen as early as late April and as late as mid-August. Optimum from mid-May to early July.

Distribution and status
Widespread and common throughout most of the British Isles, though it is absent from the Isles of Scilly.

Habitat preferences
Breeds in the still or slow-flowing waters of ponds, lakes, canals and ditches, including the acidic pools and lochans of moorland and heathland.

Description
Length *c*. 35mm.

Male (Fig 75)

Head From above, black with eyes dark red (Fig 79a). From the side, eyes dark red above with lower half reddish-brown (Fig 81a).

Thorax From above, black with antehumeral stripes bold and red (Fig 79a). In immatures the antehumeral stripes are yellow or orange (Figs 79b; 81a). From the side, a series of black and cream or white panels. On segment 3 there is a prominent black spot (Fig 81a). Legs black.

Abdomen From above, mainly red with black panels on segments 7–9 (Fig 75).

Fig. 36 Large Red Damselfly. New Forest, South Hampshire

Large Red Damselfly

From the side, similar to above but, at the base, segments 1 & 2 pale cream and orange, respectively.
Wings Clear with pterostigmata dark grey or black (Fig 81a).

Female (Fig 76)
Similar to the male but with black abdominal markings more extensive and categorised into three forms that can be identified in the field.
- **Typical form** (Fig 76)
Similar to the male but abdomen from above with a black central line that expands into a triangular spot at the rear of each of segments 2–5. Segments 6–9 mainly black. Intersegmental divisions yellow.
- **Form *fulvipes*** Stephens, 1835
Similar to the typical form but abdomen from above lacking the prominent black central line and the black triangular markings on segments 2–4.
- **Form *melanotum*** Sélys, 1876 (Fig 115)
Thorax From above, black with antehumeral stripes cream/yellow (Fig 117a).
Abdomen From above almost entirely black with a horizontal pair of tiny, cream/yellow dots at each of the segmental divisions (Fig 115).

Similar species
Male, typical female and female form *fulvipes*
See Damselfly Group 6.
Small Red Damselfly male and typical female (p 78, 79).

Female form *melanotum*
See Damselfly Group 11.
Small Red Damselfly f. *melanogastrum* (p 98, 99).

Behaviour/field tips
Often the first damselfly species to emerge in the spring, this species can sometimes be very common at its chosen sites. At the start of its flight period, and whilst not engaged in reproductive activities, it is a common sight as it hunts along sunlit woodland rides, amongst scrub and along hedgerows close to its breeding grounds.

Large Red Damselfly

CHAPTER TWO
Damselflies
Identification of the resident species

Damselfly Group 1
Wings with dark-blue patches   p 66

Fig. 37
Banded Demoiselle
male

Damselfly Group 2
Wings entirely dark   p 66

Fig. 38
Beautiful Demoiselle
male

Damselfly Group 3
Abdomen metallic bronze-green with blue 'tail'   p 68

Fig. 39
Emerald Damselfly
male

Damselfly Group 4
Abdomen metallic bronze-green without blue 'tail'   p 70

Fig. 40
Emerald Damselfly
female

Damselfly Group 5
Abdomen very pale blue or cream   p 76

Fig. 41
White-legged Damselfly
male

Damselfly Group 6
Abdomen red or red with black markings   p 77

Fig. 42
Large Red Damselfly
male

Damselfly Group 7
Abdomen black with blue 'tail'
Eyes red   p 80

Fig. 43
Small Red-eyed Damselfly male
West Norfolk

Damselfly Group 8
Abdomen black with blue 'tail'
Eyes not red   p 82

Fig. 44
Blue-tailed Damselfly male

Damselfly Group 9
Abdomen blue with
black markings   p 85

Fig. 45
Southern Damselfly male

Damselfly Group 10
Abdomen black with
blue markings   p 96

Fig. 46
Azure Damselfly female

Damselfly Group 11
Abdomen black with
red markings   p 98

Fig. 47
Small Red Damselfly female f. *melanogastrum*

Damselfly Group 12
Abdomen black with brown
or orange markings   p 101

Fig. 48
Blue-tailed Damselfly female f. *infuscans-obsoleta*

Damselfly Group 13
Abdomen black with
green markings   p 104

Fig. 49
Blue-tailed Damselfly female f. *infuscans*

Damselfly Group 1
Wings with dark-blue patches
One species: Banded Demoiselle male

Banded Demoiselle
Calopteryx splendens (Harris)

Male (Fig 50)
For species account see p 32.

Similar species
None. The male's large, dark-blue wing-patches on otherwise clear wings (Fig 50) are unique to this species.

Fig. 50 Banded Demoiselle male. West Norfolk

Damselfly Group 2
Wings entirely dark
Two species: Beautiful Demoiselle male and female, Banded Demoiselle female

Fig. 51 Beautiful Demoiselle male. South Hampshire

Fig. 52 Beautiful Demoiselle female. South Hampshire

Damselfly Group 2

Beautiful Demoiselle
Calopteryx virgo (Linn.)

Male (Fig 51)
Female (Fig 52)
For species account see p 34.

Similar species
• **Male**
None. The entirely dark-blue wings are unique to this species, see Fig 51.
• **Female**
The broad, brownish wings will usually distinguish this species, though confusion with female Banded Demoiselle can arise (Fig 53). However, it differs in the following ways:
1) Breeds in clear, fast-flowing water with a clean substrate rather than slow-moving water with a muddy sediment.
2) Wings tinged brown rather than green.

3) Pale-yellowish, dorsal stripe near the tip of abdomen clearly defined (Figs 54 & 55).

Banded Demoiselle
Calopteryx splendens (Harris)

Female (Fig 53)
For species account see p 32.

Similar species
Confusion may arise between the female of this species and that of Beautiful Demoiselle but differs in the following ways:
1) Breeds in slow-moving water over a muddy sediment rather than fast-flowing streams with a clean substrate.
2) Wings tinged green rather than brown.
3) Pale yellowish dorsal stripe near the tip of abdomen less clearly defined (Figs 54 & 55).

Fig. 53 Banded Demoiselle female West Norfolk

Fig. 54 Beautiful Demoiselle Dorsal view of female abdomen

Fig. 55 Banded Demoiselle Dorsal view of female abdomen

Damselfly Group 3
Abdomen metallic bronze-green with a blue 'tail'
Two species: Emerald Damselfly adult male, Scarce Emerald Damselfly adult male

Fig. 56 Emerald Damselfly adult male. West Norfolk

Fig. 57 Scarce Emerald Damselfly adult male. West Norfolk

Emerald Damselfly
Lestes sponsa (Hansemann)

Adult male (Fig 56)
For species account see p 30.

Similar species
Adult males of this species are easy to confuse with those of the much rarer Scarce Emerald Damselfly but differ in the following ways:
1) Blue colouration at the base of the abdomen covers the whole of the first two segments rather than all of the first and only half of the second (Fig 58a).
2) Edge of blue colouration at the tip of the abdomen usually much more clearly defined rather than fading into the green of the rest of the body. This is particularly noticeable when viewed from the side.
Note: This blue pruinesence may wear off with age (Fig 59a).
3) From above, inferior claspers at the tip of the abdomen long, narrow and straight rather than short, thick and inwardly curved (Fig 59a).
4) Pterostigmata long and narrow rather than short and broad. Not edged white (Fig 60a).

Scarce Emerald Damselfly
Lestes dryas Kirby

Adult male (Fig 57)
For species account see p 28.

Similar species
Mature males of this species may easily be confused with those of the more common Emerald Damselfly but differ in

Damselfly Group 3

the following ways:
1) Blue colouration at the base of the abdomen covers the whole of the first but only half of the second segment (Fig 58b).
2) Edge of blue colouration at the tip of the abdomen usually less clearly defined, often fading into the green of the rest of the body. This is particularly noticeable when viewed from the side. *Note*: this blue pruinesence may wear off with age (Fig 59b).
3) From above, the inferior claspers at the tip of the abdomen are short, thick and inwardly curved rather than long, narrow and straight (Fig 59b).
4) Pterostigmata short and broad rather than long and narrow. Edged white at each end (Fig 60b).

Fig. 58 Dorsal view of thorax and base of abdomen a. Emerald Damselfly b. Scarce Emerald Damselfly

Fig. 59 Dorsal view of male claspers a. Emerald Damselfly b. Scarce Emerald Damselfly

Fig. 60 Pterostigmata a. Emerald Damselfly b. Scarce Emerald Damselfly

Damselfly Group 4
Abdomen metallic bronze-green without blue 'tail'

Three species: Willow Emerald Damselfly (both sexes), Emerald Damselfly female and immature male, Scarce Emerald Damselfly female and immature male
Note:
See also the section dealing with Southern Emerald Damselfly in Chapter Five (p 230).

Fig. 61 Willow Emerald Damselfly male. East Suffolk

Fig. 62 Willow Emerald Damselfly female. East Suffolk

Fig. 63 Emerald Damselfly immature male. West Norfolk

Fig. 64 Emerald Damselfly female. West Norfolk

Damselfly Group 4

Fig. 65 Scarce Emerald Damselfly immature male. West Norfolk

Fig. 66 Scarce Emerald Damselfly female. West Norfolk

Willow Emerald Damselfly
Chalcolestes viridis
(Vander Linden)

Male (Fig 61)
Female (Fig 62)
For species account see p 26.

Similar species
Superficially similar to the female and immature male of Emerald Damselfly and Scarce Emerald Damselfly but differs in the following ways:
• **Male from immature male Emerald and Scarce Emerald Damselfly**
1) From both in having pterostigmata off-white or cream rather than black, brown or grey and with lower vein convex rather than straight (Fig 70a).
2) From Scarce Emerald Damselfly in having plain pterostigma rather than edged vertically with white (Fig 70c).
3) From both in having superior claspers white rather than dark brown or black (Fig 69a).
• **Female from female Emerald and Scarce Emerald Damselfly**
1) From both in having pterostigmata off-white or cream rather than black, brown or grey (Fig 70a).
2) From both in having a long, spur-shaped marking on the side of the thorax (Fig 68b).
3) From Emerald Damselfly in lacking the isolated green spot on the side of the thorax above the mid-leg (Fig 68b).
4) From Scarce Emerald Damselfly in having three fine, cream-coloured stripes on the top of the thorax rather than one (Fig 67a).
5) From Scarce Emerald Damselfly in having pterostigma plain rather than edged vertically white (Fig 70a).
6) Though often very difficult to see in the field, from both in having a single

Damselfly Group 4

Fig. 67 Dorsal view of thorax and base of abdomen
a. Willow Emerald Damselfly male, b. Willow Emerald Damselfly female,
c. Emerald Damselfly immature male, d. Emerald Damselfly female,
e. Scarce Emerald Damselfly immature male, f. Scarce Emerald Damselfly female

Damselfly Group 4

green bar on the top of the first abdominal segment rather than two separate spots (Fig 67a).

Emerald Damselfly
Lestes sponsa (Hansemann)

Immature male (Fig 63)
Female (Fig 64)
For species account see p 30.

Similar species
Both sexes are similar to those of Willow Emerald Damselfly, and the female and immature male of Scarce Emerald Damselfly, from which they differ in the following ways:
- **Immature male from immature male Willow Emerald Damselfly**
1) Lower vein of pterostigma straight rather than convex (Fig 70b).
2) Superior claspers very dark brown rather than white (Fig 69b).
3) From above, thorax lacks the three fine, cream-coloured stripes (Fig 67c).
- **Immature male from immature male Scarce Emerald Damselfly**
1) Pterostigmata narrow rather than broad. Not edged white at each end (Fig 70b).
2) Inferior claspers long, fine and straight rather than short, thick and inwardly curved (Fig 69b).
- **Female from female Willow Emerald Damselfly**
1) Pterostigma black rather than off-white or cream and with lower vein straight rather than convex (Fig 70b).

2) Side of thorax with green, isolated spot above mid-leg and lacking long, spur-shaped marking (Fig 68d).
- **Female from female Scarce Emerald Damselfly**
1) Generally less robust in build with slimmer abdomen.
2) Pterostigmata narrow rather than broad and plain. Not edged white at each end (Fig 70b).
3) Top of thorax with three fine, cream-coloured stripes rather than one (Fig 67d).
4) Side of thorax with green, isolated spot above the mid-leg (Fig 68d).
5) Though often difficult to see in the field, two green dorsal spots on the first abdominal segment rounded on top rather than square (Fig 67d).

Scarce Emerald Damselfly
Lestes dryas Kirby

Immature male (Fig 65)
Female (Fig 66)
For species account see p 28.

Similar species
Both are similar to Willow Emerald Damselfly, and the female and immature male of Emerald Damselfly. From these they differ in the following ways:
- **Immature male from male Willow Emerald Damselfly**
1) Pterostigma grey and edged vertically with white rather than entirely off-white or cream. Lower vein straight rather than

Damselfly Group 4

Fig. 68 Lateral view of thorax
a. Willow Emerald Damselfly male, b. Willow Emerald Damselfly female,
c. Emerald Damselfly immature male, d. Emerald Damselfly female,
e. Scarce Emerald Damselfly immature male, f. Scarce Emerald Damselfly female

Damselfly Group 4

convex (Fig 70c).
2) Superior claspers dark brown rather than white (Fig 69c).
3) From above, thorax lacks three fine, cream-coloured stripes (Fig 67e).

• **Immature male from immature male Emerald Damselfly**
1) Pterostigmata broad rather than narrow and edged vertically with white (Fig 70c).
2) Inferior claspers short, thick and inwardly curved rather than long fine and straight (Fig 69c).

• **Female from female Willow Emerald Damselfly**
1) Pterostigmata black or grey and vertically edged with white rather than off-white or cream. Lower vein straight rather than convex (Fig 70c).
2) Side of thorax lacking long, spur-shaped marking (Fig 68f).

• **Female from female Emerald Damselfly**
1) Generally more robust in build with thicker abdomen.
2) Pterostigmata broad rather than narrow and edged vertically with white (Fig 70c).
3) From the side, thorax without isolated green spot above the mid-leg (Fig 68f).
4) From above, thorax with one fine, cream-coloured stripe rather than three (Fig 67f).
5) Though difficult to see in the field, two dorsal spots on the first abdominal segment square rather than rounded at the top (Fig 67f).

 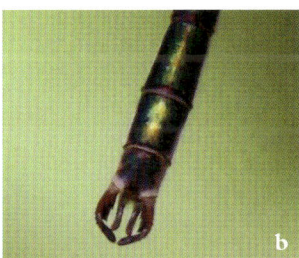 

Fig. 69 Dorsal view of male claspers a. Willow Emerald Damselfly, b. Emerald Damselfly immature, c. Scarce Emerald Damselfly immature

Fig. 70 Pterostigmata a. Willow Emerald Damselfly, b. Emerald Damselfly immature, c. Scarce Emerald Damselfly immature

Damselfly Group 5
Abdomen very pale-blue or cream-coloured
One species: White-legged Damselfly

White-legged Damselfly
Platycnemis pennipes (Pallas)

For species account see p 36.

Similar species
None. A combination of the very pale colour of the abdomen, broad, feather-like legs, amber pterostigmata and pale-blue eyes of the male (Figs 74a & b) combine to make this species unlike any other damselfly found in our region.

Fig. 71 White-legged Damselfly male. Shropshire

Fig. 72 White-legged Damselfly adult female. Shropshire

Fig. 73 White-legged Damselfly female immature form *lactea*. Shropshire

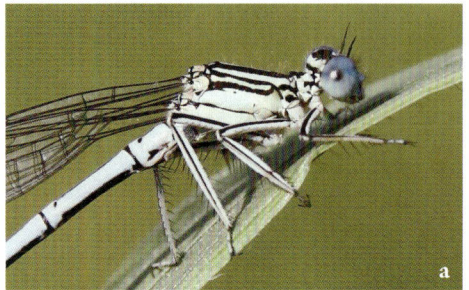

Fig. 74 White-legged Damselfly a. Lateral view of male legs and eyes, b. Pterostigmata

Damselfly Group 6
Abdomen red or red with black markings

Two species: Large Red Damselfly male and typical female, Small Red Damselfly male and typical female

Fig. 75 Large Red Damselfly male. East Norfolk

Fig. 76 Large Red Damselfly typical female. West Norfolk

Fig. 77 Small Red Damselfly male. South Hampshire

Fig. 78 Small Red Damselfly typical female. South Hampshire

Damselfly Group 6

Large Red Damselfly
Pyrrhosoma nymphula (Sulzer)

Male (Fig 75)
Female (Fig 76)
Female form *fulvipes*
For species account see p 61.

Similar species
The only other red damselfly found in our region is the Small Red Damselfly. The male, typical female and form *fulvipes* differ from male and female forms *tenellum*, *erythrogastrum* and *intermedium* of that species in the following ways:
1) Generally larger and more robust.
2) Top of thorax with broad, red or yellow antehumeral stripes (Figs 79a & b).
3) Legs black rather than red or orange (Fig 80a).
4) Pterostigma very dark grey or black rather than orange or red (Fig 81a).

Fig. 79 Dorsal view of thorax a. Large Red Damselfly male, b. Large Red Damselfly immature female, c. Small Red Damselfly male, d. Small Red Damselfly female

Damselfly Group 6

Small Red Damselfly
Ceriagrion tenellum (Villers)

Male (Fig 77)
Female (Fig 88)
For species account see p 38.

Similar species
The only other red damselfly found in our region is the Large Red Damselfly, from which the present species, in all its forms, differs in the following ways:
1) Generally smaller and more delicate in build.
2) Top of thorax lacking bold antehumeral stripes (Figs 79c & d).
3) Legs red or orange rather than black (Fig 80b).
4) Pterostigma red or orange rather than dark grey or black (Fig 81b).

Fig. 80 Lateral view of thorax and legs a. Large Red Damselfly, b. Small Red Damselfly

Fig. 81 Pterostigmata a. Large Red Damselfly, b. Small Red Damselfly

Damselfly Group 7
Abdomen black with a blue 'tail', eyes red
Two species: Red-eyed Damselfly male, Small Red-eyed Damselfly male

Fig. 82 Red-eyed Damselfly male
West Norfolk

Fig. 83 Small Red-eyed Damselfly male
West Norfolk

Red-eyed Damselfly
Erythromma najas
(Hansemann)

Adult male (Fig 82)
For species account see p 52.

Similar species
The only other species with a blue 'tail' and red eyes is the Small Red-eyed Damselfly male from which it differs in the following ways:
1) Flight period begins and ends at least two weeks earlier.
2) When perched, usually holds abdomen straight rather than curved upwards.
3) Larger and more robust in build.
4) From above, thorax lacks antehumeral stripes (Fig 84a).
5) From above, tip of the abdomen is blue without black markings (Fig 85a).
6) From the side, tip of the abdomen has only segments 9 & 10 blue (Fig 86a).
7) From the side, base of the abdomen with segment 2 black rather than blue (Fig 87a).
8) Eyes blood red rather than bright tomato red (Fig 84a).

Small Red-eyed Damselfly
Erythromma viridulum
(Charpentier)

Adult male (Fig 83)
For species account see p 54.

Similar species
The only similar species is the Red-eyed

Damselfly Group 7

Fig. 84 Dorsal view of male thorax a. Red-eyed Damselfly, b. Small Red-eyed Damselfly

Fig. 85 Dorsal view of tip of male abdomen a. Red-eyed Damselfly, b. Small Red-eyed Damselfly

Fig. 86 Lateral view of tip of male abdomen a. Red-eyed Damselfly, b. Small Red-eyed Damselfly

Damselfly Group 7

Damselfly from which it differs in the following ways:
1) Flight period usually begins, and ends, approximately two weeks later.
2) When perched, usually holds abdomen slightly curved upwards.
3) Smaller and more delicate in build.
4) From above, thorax with pale antehumeral stripes (Fig 84b).
5) From above, tip of the abdomen with black cross on segment 10 (Fig 85b).
6) From the side, tip of the abdomen with segment 8 blue rather than black (Fig 86b).
7) From the side, base of the abdomen with segment 2 blue rather than black (Fig 87b).
8) Eyes more startlingly tomato red rather than blood red (Fig 84b).

Fig. 87 Lateral view of male thorax and base of abdomen a. Red-eyed Damselfly, b. Small Red-eyed Damselfly

Damselfly Group 8
Abdomen black with blue 'tail', eyes not red

Two species: Blue-tailed Damselfly male and female forms *violacea* and *rufescens*, Scarce Blue-tailed Damselfly male

Blue-tailed Damselfly
Ischnura elegans (Vander Linden)

Adult male (Fig 88)
Immature male (Fig 89)
Female form *violacea* (Fig 90)
Female form *rufescens* (Fig 91)
For species account see p 56.

Similar species
• Male from male Scarce Blue-tailed Damselfly
The only species which may cause

Damselfly Group 8

Fig. 88 Blue-tailed Damselfly adult male. East Norfolk

Fig. 89 Blue-tailed Damselfly immature male. West Norfolk

Fig. 90 Blue-tailed Damselfly female form *violacea*. West Norfolk

Fig. 91 Blue-tailed Damselfly female form *rufescens*. East Norfolk

Fig. 92 Scarce Blue-tailed Damselfly adult male. South Hampshire

Fig. 93 Scarce Blue-tailed Damselfly immature male. South Hampshire

Damselfly Group 8

confusion is the very much more localised Scarce Blue-tailed Damselfly male from which it differs in the following ways:
1) Found in many habitat types throughout most of the British Isles rather than being highly specialized and localised (see species accounts p 28).
2) From above, tip of abdomen with segment 8 entirely blue and 9 entirely black (Fig 94a).
3) Pterostigmata long and pointed rather than short and square (Fig 95a).
• **Female forms** *violacea* **and** *rufescens*
None. The colour of the side of the thorax of forms *violacea* (violet) and *rufescens* (pink), combined with the blue 'tail', make them unmistakeable see Figs 90, 91.

Scarce Blue-tailed Damselfly
Ischnura pumilio
(Charpentier)

Adult male (Fig 92)
Immature male (Fig 93)
For species account see p 28.

Similar species
The only species which may cause confusion is the more common and widespread Blue-tailed Damselfly from which it differs in the following ways:
1) Far more localised in distribution and restricted to valley mires, flushes and seepages.
2) From above, tip of abdomen with segment 8 only partly blue and segment 9 blue with two small black markings (Fig 94b).
3) Pterostigmata short and square rather than elongate (Fig 95b).

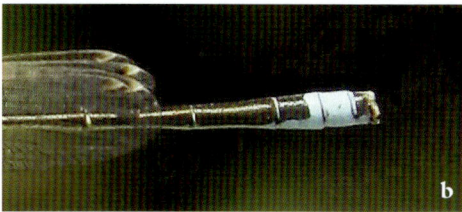

Fig. 94 Dorsal view of tip of male abdomen
a. Blue-tailed Damselfly,
b. Scarce Blue-tailed Damselfly

Fig. 95 Pterostigmata a. Blue-tailed Damselfly, b. Scarce Blue-tailed Damselfly

Damselfly Group 9
Abdomen blue with black markings

Six species: Southern Damselfly male, Northern Damselfly male, Irish Damselfly male, Azure Damselfly male and female 'blue' form, Variable Damselfly male and female 'blue' form, Common Blue Damselfly male and female 'blue' form
Note: See also Dainty Damselfly in Chapter Five (p 230).

Southern Damselfly
Coenagrion mercuriale
(Charpentier)

Male (Fig 96)
For species account see p 44.

Fig. 96 Southern Damselfly male. South Hampshire

Similar species
Of the other five species in this group that are similar in appearance, three can be discounted immediately for the following reasons:
Northern Damselfly: Geographic range does not overlap. Restricted to the Scottish Highlands.
Irish Damselfly: Geographic range does not overlap. Restricted to Ireland.
Variable Damselfly: Geographic range overlaps partly but habitat requirements completely dissimilar. Variable Damselfly prefers stagnant or very slow-flowing water with densely vegetated margins.

Confusion may arise between males of this species and males and female 'blue' forms of Azure Damselfly and Common Blue Damselfly. However, the Southern Damselfly differs from these in the following ways:
• **Male from male Azure Damselfly**
1) From above, segment 2 at base of abdomen with black marking Mercury-symbol shaped rather than 'U' shaped (Fig 101a).
2) From above, tip of abdomen with rectangular rather than crown-shaped, black marking on segment 9 (Fig 105a).
3) Head with pale post-ocular bar (Fig 101a). Absent in Azure Damselfly.
• **Male from male Common Blue Damselfly**
1) From above, segment 2 at base of abdomen with black marking Mercury-symbol shaped rather than pawn shaped (Fig 101a).
2) From above, tip of abdomen with rectangular black marking in segment 9. This segment entirely blue in Common Blue Damselfly (Fig 105a).
3) From the side, thorax with two short, black lines rather than one (Fig 102a).
• **Male from female 'blue' form of Azure Damselfly**
1) From above, segment 2 at base of

Damselfly Group 9

abdomen with black marking Mercury-symbol shaped rather than thistle shaped (Fig 101a).
2) From above, tip of abdomen with segment 8 blue rather than black (Fig 105a).
3) Head with pale post-ocular bar. Absent in Azure Damselfly (Fig 101a).
• **Male from female 'blue' form of Common Blue Damselfly**
1) As 1 and 2 above.
2) From the side, thorax with two short, black lines rather than one (Fig 102a).

Northern Damselfly
Coenagrion hastulatum (Charpentier)

Fig. 97 Northern Damselfly male. East Inverness-shire

Male (Fig 97)
For species account see p 40.

Similar species
The only two similar blue and black damselflies whose distributions are known to overlap with the present species are Common Blue and Azure, the latter having only recently begun to appear in its company. The male Northern Damselfly differs from these in the following ways:
• **Male from male Common Blue Damselfly**
1) Underside of eyes bright green rather than blue (Fig 102b).
2) From the side, thorax with two short, black stripes near wing-bases rather than one (Fig 102b).

3) From above, segment 2 at base of abdomen with black marking resembling a mushroom rather than a pawn chess piece (Fig 101b).
4) From the side, segment 2 of abdomen with an isolated black bar (Fig 102b).
5) From above, abdomen with segment 7 almost completely black rather than blue (Fig 105b).
6) From above, abdomen segment 9 with two black dots. These may be enlarged into bold spots. This segment devoid of markings in Common Blue (Fig 105b).
• **Male from male Azure Damselfly**
1) Underside of eyes bright green rather than blue (Fig 102b).
2) Head with pale post-ocular bar (Fig 101b).
3) From above, segment 2 at base of abdomen with black marking resembling a mushroom rather than the letter 'U' (Fig 101b).
4) From above, segment 3 of abdomen with large, vertical, black, missile-shaped marking (Fig 101b).

Damselfly Group 9

5) From above, segment 9 of abdomen with two black spots rather than crown-shaped marking (Fig 105b).
• **Male from female 'blue' forms of Common Blue and Azure Damselfly**
1) From above, abdomen with segment 2 predominantly blue with small, spade-shaped rather than large, black, thistle-shaped marking (Fig 101b).

Irish Damselfly
Coenagrion lunulatum
(Charpentier)

Male (Fig 98)
For species account see p 42.

Similar species
The only blue and black damselflies that live alongside this species are Azure and Variable. The Irish Damselfly differs from these in the following ways:
• **Male from male Azure Damselfly**
1) From above, base of abdomen with black marking on segment 2 crescent rather than 'U' shaped (Fig 101c).
2) From the side, base of abdomen with isolated black bar on segment 2 (Fig 102c).
3) From above, abdomen with segments 3–5 extensively black rather than blue (Fig 98).
4) From above, abdomen with two small, black dots on segment 9. Bold, black markings absent (Fig 105c).
5) Underside of abdomen, thorax and eyes green rather than blue (Fig 102c).
• **Male from male Variable Damselfly**

Fig. 98 Irish Damselfly male. Antrim

1) From above, base of abdomen with black marking on segment 2 crescent rather than wine-goblet shaped (Fig 101c).
2) From the side, base of abdomen with isolated black bar on segment 2 (Fig 102c).
3) From above, abdomen with segments 3–5 extensively black rather than blue (Fig 98).
4) From above, abdomen with two black spots on segment 9. Bold, black marking absent (Fig 105c).
5) From above, thorax with complete rather than broken, blue antehumeral stripes (Fig 101c).
6) Underside of abdomen, thorax and eyes green rather than blue (Fig 102c).
• **Male from female 'blue' forms of Azure and Variable Damselfly**
1) From above, abdomen with segment 2 predominantly blue with small, crescent-shaped rather than large, black, thistle-shaped marking (Fig 101c).
2) From above, abdomen with segments 8 & 9 almost entirely blue rather than black (Fig 105c).

Damselfly Group 9

Fig. 99 Azure Damselfly male. South Hampshire

Fig. 100 Azure Damselfly female 'blue' form. West Norfolk

Azure Damselfly
Coenagrion puella (Linn.)

Male (Fig 99)
Female 'blue' form (Fig 100)
For species account see p 46.

Similar species
Because of its widespread distribution, the Azure Damselfly might be encountered in company with any of the others in this group, though it appears that the species has only recently been found within the range of Northern Damselfly. Only two, Variable and Common Blue, have female 'blue' forms similar in appearance to that of the present species. The Azure Damselfly differs from those of the other species in the group in the following ways:
• **Male from male Common Blue Damselfly**
1) From above, base of abdomen with black, 'U'-shaped rather than pawn-shaped marking on segment 2 (Fig 101d).
2) From above, tip of abdomen with bold, black, crown-shaped marking on segment 9 (Fig 105d).
3) From the side, thorax with two black stripes near wing bases rather than one (Fig 102d).
4) Head without pale post-ocular bar (Fig 101d).
• **Male from male Southern Damselfly**
1) Southern Damselfly with highly restricted distribution (see species account p 44).
2) From above, base of abdomen with black, 'U'-shaped rather than Mercury-symbol-shaped marking on segment 2 (Fig 101d).
3) From above, abdomen with black rectangle in lower third of segment 3. Rest of segment plain blue (Fig 101d).
4) From above, abdomen with black, crown-shaped marking on segment 9 rather than a solid black rectangle (Fig 105d).
5) Head without pale-blue post-ocular bar (Fig 101d).
• **Male from male Northern Damselfly**

Damselfly Group 9

1) Northern Damselfly restricted to Scottish Highlands. Absent from England, Wales and Ireland.
2) From above, abdomen with black, 'U'-shaped rather than spade-shaped marking on segment 2 (Fig 101d).
3) From above, abdomen with black rectangle in lower third of segment 3. Rest of segment plain blue (Fig 101d).
4) From above, abdomen with bold, black, crown-shaped marking on segment 9 rather than two black spots (Fig 105d).
5) From the side, abdomen without isolated black bar on segment 2 (Fig 102d).
6) From the side, lower half of eyes and thorax blue rather than green (Fig 102d).
7) Head without pale post-ocular bar (Fig 101d).

• **Male from male Irish Damselfly**
1) Irish Damselfly restricted to Ireland. Absent from England, Wales and Scotland.
2) From above, abdomen with black, 'U'-shaped rather than crescent-shaped marking on segment 2 (Fig 101d).
3) From above, abdomen with only lower third rather than most of segment 3 black (Fig 101d).
4) From above, abdomen with black,

Fig. 101 Dorsal view of male head, thorax and base of abdomen
a. Southern Damselfly, b. Northern Damselfly, c. Irish Damselfly,
d. Azure Damselfly, e. Variable Damselfly, f. Common Blue Damselfly

Damselfly Group 9

crown-shaped marking on segment 9 rather than two small dots (Fig 105d).
5) From the side, abdomen without isolated black bar on segment 2 (Fig 102d).
6) From the side, lower half of eyes and thorax blue rather than green (Fig 102d).
• **Male from male Variable Damselfly**
1) Variable Damselfly very much more restricted in distribution (see species account p 48).
2) From above, abdomen with black, 'U'-shaped rather than wine-goblet-shaped marking on segment 2 (Fig 101d).
3) From above, abdomen with black, crown-shaped marking on segment 9 rather than solid rectangle (Fig 105d).
4) Thorax with complete rather than broken antehumeral stripes (Fig 101d).
5) Head without pale post-ocular bar (Fig 101d).
• **Female from female 'blue' form of Common Blue Damselfly**
1) From above, abdomen lacking black, bomb-shaped markings on central segments (Fig 100).
2) From the side, thorax with two rather than one black stripes near wing bases (Fig 109a).
3) Head without pale post-ocular bar (Fig 108a).

Fig. 102 Lateral view of male head, thorax and base of abdomen
a. Southern Damselfly, b. Northern Damselfly, c. Irish Damselfly © Graham Sherwin,
d. Azure Damselfly, e. Variable Damselfly, f. Common Blue Damselfly

Damselfly Group 9

- **Female from female 'blue' form of Variable Damselfly**
Head without pale post-ocular bar (Fig 108a).

Variable Damselfly
Coenagrion pulchellum
(Vander Linden)

Male (Fig 103)
Female 'blue' form (Fig 104)
For species account see p 48.

Similar species
This species is unlikely to be encountered alongside Northern Damselfly as the geographical ranges of the two do not overlap. Although the ranges of the Variable and Southern Damselfly overlap slightly, their different habitat requirements probably preclude coexistence. However, confusion is likely with the other three species in the group from which it differs in the following ways:
- **Male from male Irish Damselfly**
1) Irish Damselfly restricted to Ireland. Absent from elsewhere in the British Isles.
2) From above, base of abdomen with black marking on segment 2 resembling a wine goblet rather than a crescent (Fig 101e).
3) From above, abdomen with segment 3 mainly blue rather than with bold, black marking (Fig 101e).
4) From above, abdomen segment 9 with solid black, rectangular marking rather than two black dots (Fig 105e).
5) From above, thorax with antehumeral

Fig. 103 Variable Damselfly male. Cambridgeshire

Fig. 104 Variable Damselfly female 'blue' form. East Norfolk

stripes broken to form an exclamation-shaped marking rather than being complete (Fig 101e).
6) From the side, base of abdomen lacking isolated black bar on segment 2 (Fig 102e).
7) From the side, eyes and thorax mainly blue below rather than green (Fig 102e).
8) Head with pale post-ocular bar (Fig 101e).

Damselfly Group 9

- **Male from male Azure Damselfly**
1) From above, segment 2 of abdomen with black marking resembling a wine goblet rather than a 'U' (Fig 101e).
2) From above, abdomen with solid black rectangular rather than crown-shaped marking on segment 9 (Fig 105e).
3) From above, thorax with broken antehumeral stripes forming an exclamation-shaped marking rather than being complete (Fig 101e).
4) Head with pale post-ocular bar (Fig 101e).
- **Male from male Common Blue Damselfly**
1) From above, segment 2 of abdomen with black marking resembling a wine goblet rather than a chess pawn (Fig 101e).
2) From above, abdomen with segment 9 almost wholly black rather than blue (Fig 105e).
3) From above, thorax with broken antehumeral stripes forming an exclamation-shaped marking rather than being complete (Fig 101e).
4) From the side, thorax with two short, black stripes near wing bases rather than one (Fig 102e).

Fig. 105 Dorsal view of tip of male abdomen
a. Southern Damselfly, b. Northern Damselfly, c. Irish Damselfly,
d. Azure Damselfly, e. Variable Damselfly, f. Common Blue Damselfly

Damselfly Group 9

The female 'blue' form is generally darker in appearance than any of the males of other species in this group and can be distinguished from them by the black, thistle-shaped marking at the base of the abdomen on the top of segment 2. However, confusion could arise between this species and the female 'blue' forms of Azure and Common Blue Damselfly. It differs from these in the following ways:

• **Female from female 'blue' form of Azure Damselfly**
Head from above with pale post-ocular bar (Fig 108b).

• **Female from female 'blue' form of Common Blue Damselfly**
Thorax from the side with two short, black stripes near wing-bases rather than one (Fig 109b).

Common Blue Damselfly
Enallagma cyathigerum (Charpentier)

Fig. 106 Common Blue Damselfly male. East Norfolk

Fig. 107 Common Blue Damselfly female 'blue' form. West Norfolk

Male (Fig 106)
Female 'blue' form (Fig 107)
For species account see p 50.

Similar species
Because of its widespread distribution, the Common Blue might be encountered in company with any of the other species in this group. The females are slightly more straightforward to identify as only the Variable and Azure have 'blue' forms similar in appearance to the present species. The Common Blue differs from the other damselflies in the group in the following ways:

• **Male from male Southern Damselfly**
1) From above, base of abdomen with black marking on segment 2 resembling a chess pawn rather than the sign for Mercury (Fig 101f).
2) From above, abdomen segment 3 lacking 'missile'-shaped marking (Fig 101f).
3) From above, abdomen segment 9

Damselfly Group 9

wholly blue rather than with large, black rectangle (Fig 105f).
4) From the side, thorax with one short, black stripe near base of wings rather than two (Fig 102f).

• **Male from male Northern Damselfly**
1) From above, base of abdomen with black marking resembling a chess pawn rather than a mushroom (Fig 101f).
2) From above, abdomen segment 3 lacking long, black, triangular marking (Fig 101f).
3) From above, abdomen segment 9 wholly blue rather than with two black spots (Fig 105f).

4) From the side, thorax with one short, black stripe near the base of the wings rather than two (Fig 102f).
5) From the side, thorax lacking isolated black bar (Fig 102f).
6) From the side, underside of thorax and eyes blue rather than green (Fig 102f).

• **Male from male Irish Damselfly**
1) From above, base of abdomen with black marking resembling a chess pawn rather than a crescent (Fig 101f).
2) From above, abdomen segment 3 mainly blue rather than black (Fig 101f).
3) From above, abdomen segment 9

Fig. 108 Dorsal view of female head, thorax and base of abdomen
a. Azure Damselfly 'blue' form, b. Variable Damselfly 'blue' form,
c. Variable Damselfly 'blue' form, d. Common Blue Damselfly 'blue' form

Damselfly Group 9

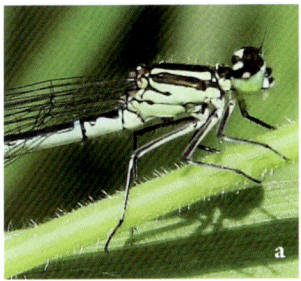  

Fig. 109 Lateral view of female thorax a. Azure Damselfly 'blue' form, b. Variable Damselfly 'blue' form, c. Common Blue Damselfly 'blue' form

wholly blue rather than with two black spots (Fig 105f).
4) From the side, thorax and eyes blue below rather than green (Fig 102f).
5) From the side, thorax with one black stripe near the base of the wings rather than two (Fig 102f).
6) From the side, thorax lacking isolated black bar (Fig 102f).

- **Male from male Azure Damselfly**

1) From above, base of abdomen with black marking resembling a chess pawn rather than a drinking tumbler (Fig 101f).
2) From above, abdomen segment 9 wholly blue without bold, black markings (Fig 105f).
3) From the side, thorax with one short, black stripe near the base of the wings rather than two (Fig 102f).
4) Head with pale post-ocular bar (Fig 101f).

- **Male from male Variable Damselfly**

1) From above, base of abdomen with black marking on segment 2 resembling a chess pawn rather than a wine goblet (Fig 101f).
2) From above, abdomen segment 9 almost entirely blue rather than black (Fig 105f).
3) From the side, abdomen with lower half of segment 3 blue rather than black (Fig 102f).
4) From the side, thorax with one short, black stripe near the base of the wings rather than two (Fig 102f).
5) From above, thorax with antehumeral stripes complete rather than broken to form an exclamation marking (Fig 101f).

- **Male from female 'blue' forms of Variable and Azure Damselfly**

From above, abdomen segment 1with black marking on segment 2 resembling a chess pawn rather than a thistle (Fig 101f).

- **Female from female 'blue' form of Azure Damselfly**

1) From above, central segments of abdomen each with a black marking resembling a torpedo or bomb (Fig 107).
2) From the side, thorax with one short, black stripe near base of wings rather than two (Fig 109c).
3) From above, head with pale post-ocular bar (Fig 108d).

- **Female from female 'blue' form of Variable Damselfly**

See 1and 2 above.

Damselfly Group 10
Abdomen black with blue markings

Three species: Variable Damselfly female 'dark' form, Azure Damselfly female 'blue' form, Small Red-eyed Damselfly female

Variable Damselfly
Coenagrion pulchellum
(Vander Linden)

Female 'dark' form (Fig 110)
For species account see p 48.

Similar species
Could easily be confused with the females of the other two species in this group but differs from these in the following ways:
- **From Azure Damselfly 'blue' form**
From above, head with pale post-ocular bar (Fig 113a).
- **From Small Red-eyed Damselfly**
1) From above, thorax with antehumeral stripes blue rather than green (Fig 113a).
2) From above, head with conspicuous

Fig. 110 Variable Damselfly female 'dark' form. East Norfolk

Fig. 111 Azure Damselfly female 'blue' form. East Norfolk

Fig. 112 Small Red-eyed Damselfly female West Norfolk

Damselfly Group 10

post-ocular spots (Fig 113a).
3) From the side, eyes black above and blue below rather than brown above and green below (Fig 114a).

Azure Damselfly
Coenagrion puella (Linn.)

Female 'blue' form (Fig 111)
For species account see p 46.

Similar species
Could easily be confused with the females of the other two species in this group but differs from these in the following ways:
• **From Variable Damselfly 'dark' form**
From above, head without pale post-ocular bar (Fig 113b).
• **From Small Red-eyed Damselfly**
1) From above, thorax with antehumeral stripes blue rather than green (Fig 113b).
2) From above, head with conspicuous post-ocular spots (Fig 113b).
3) From the side, eyes black above and blue below rather than brown above and green below (Fig 114b).

Small Red-eyed Damselfly
Erythromma viridulum (Charpentier)

Female (Fig 112)
For species account see p 54.

Similar species
Could be confused with the females of

Fig. 113 Dorsal view of female head and thorax a. Variable Damselfly 'dark' form, b. Azure Damselfly 'blue' form, c. Small Red-eyed Damselfly

Damselfly Group 10

Fig. 114 Lateral view of female head and thorax a. Variable Damselfly 'dark' form, b. Azure Damselfly 'blue' form, c. Small Red-eyed Damselfly

the other two species in this group but differs from both in the following ways:
1) From above, thorax with antehumeral stripes green rather than blue (Fig 113c).
2) From above, head without conspicuous post-ocular spots (Fig 113c).
3) From the side, eyes brown above and green below rather than black above and blue below (Fig 114c).

Damselfly Group 11
Abdomen black with red markings

Two species: Large Red Damselfly female form *melanotum*, Small Red Damselfly female form *melanogastrum*

Large Red Damselfly
Pyrrhosoma nymphula (Sulzer)

Female form *melanotum* (Fig 115)
For species account see p 61.

Similar species
May be confused with the female form *melanogastrum* of Small Red Damselfly but differs in the following ways:

1) Far more widespread and found throughout most of the British Isles rather than being very localised (see species account p 38).
2) Viewed from above, abdomen with red and yellow spots rather than almost entirely bronze-black (Fig 115).
3) From above, thorax with bold, yellow antehumeral stripes (Fig 117a).
4) Legs black rather than brown or red (Fig 118a).
5) Pterostigmata black rather than brown or red (Fig 119a).

Damselfly Group 11

Small Red Damselfly
Ceriagrion tenellum (Villers)

Female form *melanogastrum* (Fig 116)
For species account see p 38.

Similar species
Differs from the female form *melanotum* of Large Red Damselfly in the following ways:
1) Far more restricted in its geographical range and absent from large areas of the British Isles (see species account p 38).

Fig. 115 Large Red Damselfly form *melanotum*, South Hampshire

Fig. 116 Small Red Damselfly form *melanogastrum*, South Hampshire

Fig. 117 Dorsal view of thorax
a. Large Red Damselfly form *melanotum*, b. Small Red Damselfly form *melanogastrum*

Damselfly Group 11

2) From above, abdomen almost entirely bronze-black with only tiny, yellow spots (Fig 117b).
3) From above, thorax with antehumeral stripes very fine or entirely absent (Fig 117b).
4) Legs brown or red rather than black (Fig 118b).
5) Pterostigmata brown or red rather than black (Fig 119b).

Fig. 118 Lateral view of thorax and legs
a. Large Red Damselfly form *melanotum*, b. Small Red Damselfly form *melanogastrum*

Fig. 119 Pterostigmata
a. Large Red Damselfly form *melanotum*, b. Small Red Damselfly form *melanogastrum*

Damselfly Group 12
Abdomen black with pale-brown or orange markings

Three species: Common Blue Damselfly female 'brown' form, Scarce Blue-tailed Damselfly female form *aurantiaca*, Blue-tailed Damselfly female form *infuscans-obsoleta*

Fig. 120 Common Blue Damselfly female 'brown' form. East Suffolk © Kerry Robinson

Common Blue Damselfly
Enallagma cyathigerum (Charpentier)

Female 'brown' form (Fig 120)
For species account see p 50.

Similar species
The row of dark-brown, bomb-shaped markings along the upper surface of the abdomen should be diagnostic but the female also differs from the females of the other two species in this group in the following ways:
• **From both species**
1) Head with pale post-ocular bar (Fig 123a).
2) Post-ocular spots elongate rather than triangular (Fig 122a).
3) From above, thorax with broad, pale antehumeral stripes (Fig 122a).
4) From the side, thorax with bold, black horizontal stripe below antehumeral stripe (Fig 123a).
5) Pterostigmata evenly pale rather than bicoloured (Fig 125a).
• **From Scarce Blue-tailed Damselfly form *aurantiaca***
As above but also:
From above, base of abdomen with large, black marking on segment 2 rather than plain (Fig 122a).

Scarce Blue-tailed Damselfly
Ischnura pumilio (Charpentier)

Female form *aurantiaca* (Fig 121)
For species account see p 28.

Similar species
The fluorescent orange colour of the

Fig. 121 Scarce Blue-tailed Damselfly female form *aurantiaca*. South Hampshire

Damselfly Group 12

early immature phase is unmistakeable, but as this darkens with maturity, confusion may arise with the females of the other two species in the group. It differs from them in the following ways:

• **From both species**
From above, base of abdomen with segment 2 plain orange or pale brown rather than with a large, black marking (Fig 122b).

• **From Common Blue Damselfly 'brown' form**
1) Head without pale post-ocular bar (Fig 123b).
2) Post-ocular spots triangular rather than elongate (Fig 122b).
3) From above, thorax without pale antehumeral stripes (Fig 123b).
4) From the side, thorax plain. Dark horizontal stripe absent (Fig 123b).
5) From above, abdomen with central segments almost entirely black. Row of bomb-shaped markings absent (Fig 121).
6) Pterostigmata bicoloured rather than plain (Fig 125b).

• **From Blue-tailed Damselfly form**

Fig. 122 Dorsal view of head, thorax and base of abdomen
a. Common Blue Damselfly female 'brown' form, b. Scarce Blue-tailed Damselfly female form *aurantiaca*, c. Blue-tailed Damselfly female form *infuscans-obsoleta*

 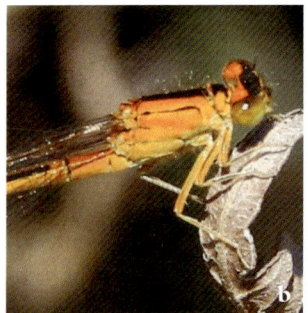 

Fig. 123 Lateral view of head, thorax and base of abdomen
a. Common Blue Damselfly female 'brown' form, b. Scarce Blue-tailed Damselfly female form *aurantiaca*, c. Blue-tailed Damselfly female form *infuscans-obsoleta*

Damselfly Group 12

infuscans-obsoleta
1) From above, base of abdomen with segment 2 plain orange or pale brown rather than black (Fig 122b).
2) Pterostigmata short and square rather than long and pointed (Fig 125b).

Blue-tailed Damselfly
Ischnura elegans
(Vander Linden)

Fig. 124 Blue-tailed Damselfly female form *infuscans-obsoleta*. South Hampshire

Female form *infuscans-obsoleta* (Fig 124)
For species account see p 56.

Similar species
Differs from the females of the other two species in this group in the following ways:
• **From Common Blue Damselfly 'brown' form**
1) Head without pale post-ocular bar (Fig 123c).
2) Post-ocular spots triangular rather than elongate (Fig 122c).
3) From above, thorax lacking pale antehumeral stripes (Fig 122c).
4) From the side, thorax without black horizontal stripe (Fig 123c).
5) From above, abdomen with central segments almost entirely black. Row of bomb-shaped markings absent (Fig 124).
6) Pterostigmata bicoloured rather than plain (Fig 125c).
• **From Scarce Blue-tailed Damselfly form *aurantiaca***
1) From above, base of abdomen with segment 2 mainly black rather than orange or pale brown (Fig 122c).
2) Pterostigmata long and pointed rather than short and square (Fig 125c).

Fig. 125 Pterostigmata a. Common Blue Damselfly female 'brown' form, b. Scarce Blue-tailed Damselfly female f. *aurantiaca*, c. Blue-tailed Damselfly female f. *infuscans-obsoleta*

Damselfly Group 13
Abdomen black with green markings

Seven species: Northern Damselfly female, Irish Damselfly female, Southern Damselfly female, Azure Damselfly female 'green' form, Scarce Blue-tailed Damselfly adult female, Blue-tailed Damselfly female form *infuscans*, Red-eyed Damselfly female and immature male

Fig. 126 Northern Damselfly female. East Inverness-shire

Northern Damselfly
Coenagrion hastulatum (Charpentier)

Female (Fig 126)
For species account see p 40.

Similar species
Within this group, the Northern Damselfly shares its geographical range with only Azure Damselfly and Blue-tailed Damselfly (see species account p 46, 56). The female differs from the females of both of these in the following ways:

- **From Azure Damselfly 'green' form**
1) From above, head with pale post-ocular bar (Fig 132a).
2) From the side, lower half of eyes pure green rather than tinged with blue (Fig 133a).
- **From female Blue-tailed Damselfly form *infuscans***
1) Post-ocular spots elongate rather than triangular (Fig 132a).
2) Head with pale post-ocular bar (Fig 132a).
3) Lower half of eyes pure green rather than tinged with brown (Fig 133a).
4) From the side, thorax with two bold rather than small, faint, black stripes at wing-bases (Fig 133a).
5) From above, abdomen with segment 8 mainly black rather than brown (Fig 136).

Irish Damselfly
Coenagrion lunulatum (Charpentier)

Female (Fig 127)
For species account see p 42.

Similar species
Although the distribution of the Scarce Blue-tailed Damselfly partly overlaps, its habitat preferences are so dissimilar to those of the present species that they are

Damselfly Group 13

very unlikely to be found flying together. Confusion could occur with Azure Damselfly and Blue-tailed Damselfly, but the female differs from the females of both of these in the following ways:

• **From Azure Damselfly 'green' form**
1) Head with post-ocular spots shorter, more triangular and well separated (Fig 132b).
2) From the side, lower half of eyes pure green rather than tinged with blue (Fig 133b).
3) From above, thorax with antehumeral stripes tinged strongly with brown rather than pure green (Fig 132b).
4) From the side, upper half of thorax tinged strongly with brown rather than pure green (Fig 133b).
5) From above, tip of abdomen with more extensive pale markings on segment 8 (Fig 127).

• **From female Blue-tailed Damselfly form *infuscans***
1) Head with post-ocular spots more elongate (Fig 132b).
2) From the side, lower half of eyes pure green rather than tinged with brown (Fig 133b).
3) From the side, thorax with two bold rather than small, faint, black stripes (Fig 133b).
4) From above, abdomen with segment 8 mainly black rather than brown (Fig 136).

Southern Damselfly
Coenagrion mercuriale (Charpentier)

Female (Fig 128)
For species account see p 44.

Similar species
In this group, the only species with which the Southern Damselfly overlaps in distribution and habitat requirements

Fig. 127 Irish Damselfly female. Antrim

Fig. 128 Southern Damselfly female. South Hampshire

Damselfly Group 13

are Azure Damselfly, Scarce Blue-tailed Damselfly and Blue-tailed Damselfly f. *infuscans*. The female differs from the females of these in the following ways:
- **From Azure Damselfly 'green' form**
1) Head with triangular post-ocular spots short and more equilateral in shape rather than elongate (Fig 132c).
2) Head with pale post-ocular bar (Fig 132c).
- **From Scarce Blue-tailed Damselfly**
1) Head with post-ocular spots pale green and conspicuous (Fig 132c).
2) Head with pale post-ocular bar (Fig 132c).
3) Eyes with lower half pale bluish rather than pure green (Fig 133c).
4) From above, thorax with bold antehumeral stripes (Fig 132c).
5) From the side, thorax with two bold stripes near wing-bases (Fig 133c).
- **From Blue-tailed Damselfly form** *infuscans*
1) Head with pale post-ocular bar (Fig 132c).
2) Eyes with lower half pale bluish-green rather than pure green or tinged brown (Fig 133c).
3) From the side, thorax with two bold rather than inconspicuous black stripes at wing-bases (Fig 133c).
4) From above, abdomen with segment 8 mainly black rather than pale brown (Fig 136).

Azure Damselfly
Coenagrion puella (Linn.)

Female 'green' form (Fig 129)
For species account see p 46.

Similar species
This widespread species may be encountered in company with any of the others in this group but the female differs from them in the following ways:
- **From female Northern Damselfly**
1) Northern Damselfly restricted to the Scottish Highlands (see species account p 40) and therefore absent from most of this species' geographical range.
2) Head lacking pale post-ocular bar (Fig 132d).
3) Eyes with lower half pale greenish-blue rather than pure green (Fig 133d).
- **From female Irish Damselfly**
1) Irish Damselfly restricted to small parts of Ireland (see species account p 42) and therefore absent from most of

Fig. 129 Azure Damselfly female 'green' form. East Norfolk

Damselfly Group 13

this species' geographical range.
2) From above, thorax with antehumeral stripes pure blue rather than tinged with brown (Fig 132d).
3) From the side, thorax with upper half bluish-green rather than tinged with brown (Fig 133d).
4) Eyes with lower half tinged blue rather than pure green (Fig 133d).
• **From female Southern Damselfly**
Head lacking pale post-ocular bar (Fig 132d).
• **From female Scarce Blue-tailed Damselfly**
1) Head from above with conspicuous post-ocular spots (Fig 132d).
2) Head with post-ocular spots elongate rather than equilateral (Fig 132d).
3) Eyes with lower half pale bluish rather than pure green (Fig 133d).
4) From above, thorax with bold antehumeral stripes (Fig 132d).
5) From the side, thorax with two bold, black stripes at wing-bases (Fig 133d).
• **From female Blue-tailed Damselfly form *infuscans***
1) Head with post-ocular spots elongate rather than equilateral (Fig 132d).
2) Eyes with lower half pale bluish-green rather than tinged brown (Fig 133d).
3) Thorax from side with two bold rather than inconspicuous black stripes at wing-bases (Fig 133d).
4) Abdomen from above with segment 8 mainly black rather than pale brown (Fig 136).
• **From female and immature male Red-eyed Damselfly**
1) Head with post-ocular spots (absent on Red-eyed Damselfly) (Fig 132d).

2) From above, thorax with complete and conspicuous antehumeral stripes (Fig 132d).

Scarce Blue-tailed Damselfly
Ischnura pumilio (Charpentier)

Adult female (Fig 130)
For species account see p 58.

Similar species
Because of their isolated distributions, this species is unlikely to be found flying in company with Northern Damselfly (restricted to Scotland) or Irish Damselfly (restricted to Ireland). Different habitat requirements also make confusion with Red-eyed Damselfly unlikely. It differs from the females of the

Fig. 130 Scarce Blue-tailed Damselfly adult female. South Hampshire

Damselfly Group 13

other three species in this group in the following ways:
- **From female Southern Damselfly**
1) Head without pale post-ocular bar (Fig 132e).
2) Eyes with lower half pure green rather bluish-green (Fig 133e).
3) From above, thorax without antehumeral stripes (Fig 132e).
4) From the side, thorax without black lines at wing-bases (Fig 133e).
- **From Azure Damselfly 'green' form**
As above but also:
Head with triangular post-ocular spots equilateral in shape rather than elongate (Fig 132e).
- **From Blue-tailed Damselfly form** *infuscans*
1) From above, thorax without antehumeral stripes (Fig 132e).
2) From above, abdomen with segment 8 black rather than pale brown (Fig 136).
3) Pterostigmata shorter and squarer (Fig 135a).

Blue-tailed Damselfly
Ischnura elegans
(Vander Linden)

Female form *infuscans* (Fig 131)
For species account see p 56.

Similar species
This form of Blue-tailed Damselfly may be found flying alongside all of the other species in this group. The female differs

Fig. 131 Blue-tailed Damselfly form *infuscans*. **South Hampshire**

from them in the following ways:
- **From female Northern Damselfly**
1) Northern Damselfly restricted to the Scottish Highlands and absent from most of this species' range (see species account p 40).
2) Triangular post-ocular spots equilateral in shape rather than elongate (Fig 132f).
3) Head without pale post-ocular bar (Fig 132f).
4) From the side, thorax with two black stripes near wing-bases small and inconspicuous (Fig 133f).
5) From above, abdomen with segment 8 pale brown rather than black (Fig 136).
6) Pterostigmata bicoloured (Figs 135b; 136).
- **From female Irish Damselfly**
1) Irish Damselfly restricted to parts of Ireland and absent from most of this species' range (see species account p 42).
2) Triangular post-ocular spots equilateral in shape rather than elongate (Fig 132f).

Damselfly Group 13

3) From the side, thorax with two black stripes near wing-bases small and inconspicuous (Fig 133f).
4) From the side, thorax with upper half pure green rather than tinged brown (Fig 133f).
5) From above, abdomen with segment 8 pale brown rather than black (Fig 136).
6) Pterostigmata bicoloured (Figs 135b; 136).
• **From female Southern Damselfly**
1) Southern Damselfly restricted to small areas of southern England, southern Wales and Anglesey, and absent from most of this species' range (see species account p 44).
2) Head without pale post-ocular bar (Fig 132f).

Fig. 132 Dorsal view of thorax
a. Northern Damselfly female, b. Irish Damselfly female, c. Southern Damselfly female, d. Azure Damselfly female 'green' form, e. Scarce Blue-tailed Damselfly adult female, f. Blue-tailed Damselfly form *infuscans*, g. Red-eyed Damselfly female

Damselfly Group 13

3) From the side, eyes tinged brownish rather than blue (Fig 133f).
4) From the side, thorax with two black stripes near wing-bases small and inconspicuous (Fig 133f).
5) From above, abdomen with segment 8 pale brown rather than black (Fig 136).
6) Pterostigmata bicoloured (Figs 135b; 136).
- **From female Azure Damselfly 'green' form**
1) Triangular post-ocular spots equilateral in shape rather than elongate (Fig 132f).
2) From the side, eyes tinged brownish rather than blue (Fig 133f).
3) From the side, thorax with two black stripes near wing-bases small and inconspicuous (Fig 133f).

Fig. 133 Lateral view of thorax
a. Northern Damselfly female, b. Irish Damselfly female, c. Southern Damselfly female, d. Azure Damselfly female 'green' form, e. Scarce Blue-tailed Damselfly adult female, f. Blue-tailed Damselfly form *infuscans*, g. Red-eyed Damselfly female

Damselfly Group 13

4) From above, abdomen with segment 8 pale brown rather than black (Fig 136).
5) Pterostigmata bicoloured (Figs 135b; 136).
• **From female Scarce Blue-tailed Damselfly**
1) Scarce Blue-tailed Damselfly with much more restricted distribution and absent from most of this species' range (see species account p 58).
2) Head without pale post-ocular bar (Fig 132f).
3) From above, thorax with bold ante-humeral stripes (Fig 132f).
4) From above, abdomen with segment 8 pale brown rather than black (Fig 136).
5) Pterostigma longer and more pointed (Figs 135b; 136).
• **From female and immature Red-eyed Damselfly**
1) Red-eyed Damselfly more restricted in its distribution and absent from Scotland, Ireland, most of Wales and northern England (see species account p 52).
2) Head without post-ocular spots (Fig 132f).
3) From above, thorax with antehumeral stripes short and yellow (Fig 132f).
4) From the side, thorax with very short and inconspicuous rather than bold, black stripes at wing-bases (Fig 133f).
5) From above, abdomen with segment 8 pale brown rather than black (Fig 136).
6) Pterostigmata bicoloured (Figs 135b; 136).

Fig. 134 Red-eyed Damselfly female. West Norfolk

Red-eyed Damselfly
Erythromma najas (Hansemann)

Female (Fig 134) **and immature male**
For species account see p 52.

Similar species
The absence of this species from Ireland and Scotland prevents it being confused with the Irish Damselfly and Northern Damselfly, respectively. Further, its preference for still or very slow-flowing water with lush surface vegetation or algal mats makes it unlikely to be found alongside Scarce Blue-tailed Damselfly (see p 58) or Southern Damselfly (see p 44). It differs from the other two species in this group in the following ways:
• **Female and immature male from female 'green' form Azure Damselfly**
1) Absent from Scotland, Ireland and

Damselfly Group 13

Fig. 135 Pterostigmata
a. Scarce Blue-tailed Damselfly, b. Blue-tailed Damselfly form *infuscans*

most of Wales and northern England.
2) Head without post-ocular spots (Fig 132g).
3) From the side, eyes with lower half brownish-green rather than bluish-green (Fig 133g).
4) From above, thorax with antehumeral stripes short or absent rather than long and bold (Fig 132g).
5) From the side, thorax with lower of the two black stripes at the wing bases reaching hind leg (Fig 133g).

• **Female and immature male from female Blue-tailed Damselfly form** *infuscans*
1) Generally more robust in build.
2) From above, thorax lacking bold antehumeral stripes (Fig 132g).
3) From the side, thorax with two black stripes long rather than short (Fig 133g).
4) From above, abdomen with segmental divisions blue (Fig 134).
5) From above, abdomen with segment 8 mainly black rather than brown (see Fig 136).

Fig. 136 Dorsal view of tip of abdomen of Blue-tailed Damselfly form *infuscans*

CHAPTER THREE
Species accounts of the resident Dragonflies

Azure Hawker *Aeshna caerulea*	114
Southern Hawker *Aeshna cyanea*	116
Brown Hawker *Aeshna grandis*	119
Common Hawker *Aeshna juncea*	121
Migrant Hawker *Aeshna mixta*	124
Norfolk Hawker *Anaciaeschna isosceles*	127
Emperor Dragonfly *Anax imperator*	129
Hairy Dragonfly *Brachytron pratense*	131
Club-tailed Dragonfly *Gomphus vulgatissimus*	134
Golden-ringed Dragonfly *Cordulegaster boltonii*	137
Downy Emerald Dragonfly *Cordulia aenea*	138
Northern Emerald Dragonfly *Somatochlora arctica*	141
Brilliant Emerald Dragonfly *Somatochlora metallica*	144
White-faced Darter *Leucorrhinia dubia*	146
Broad-bodied Chaser *Libellula depressa*	148
Scarce Chaser *Libellula fulva*	150
Four-spotted Chaser *Libellula quadrimaculata*	152
Black-tailed Skimmer *Orthetrum cancellatum*	154
Keeled Skimmer *Orthetrum coerulescens*	157
Black Darter *Sympetrum danae*	160
Red-veined Darter *Sympetrum fonscolombii*	163
Common Darter *Sympetrum striolatum*	165
Ruddy Darter *Sympetrum sanguineum*	168

Azure Hawker
Aeshna caerulea (Ström, 1783)

Fig. 137 Azure Hawker.
Beinn Eighe, West Ross. © Graham Sherwin

Flight period
Usually early June to late July. Because of vagaries of the climate in the Scottish Highlands, the flight period may be advanced, beginning in early May, or delayed, and continuing into early August. Optimum from mid-June to mid-July.

Distribution and status
When reading this summary, one must be aware that this species is very localised and often inhabits sites that are difficult to access and at the mercy of extremely inclement weather. This is very restrictive to field recording and there is therefore much scope for the discovery of 'new' colonies. The following should therefore be regarded only as a broad guide.
Scotland: The Azure Hawker is restricted in its range to the upland areas centred in the counties of West and East Inverness-shire and West Ross. There is also an outpost in south-west Scotland centred in the north of Kirkcudbrightshire. Scattered records and reports of apparently 'lost' colonies are frequent from the peripheries of the afore-mentioned focal areas.
England, Wales and Ireland: Absent

Habitat preferences
The Azure Hawker inhabits shallow *Sphagnum*-rich bog pools on mountain plateaux and raised peat bogs in poorly drained valleys. Many of the breeding sites are remote and very difficult to traverse.

Description
Length *c*. 60mm; wingspan *c*. 85mm.
Male (Fig 204)
Head From above, frons pale yellowish-blue with a black, T-shaped, central marking. Eyes slate-grey tinged with blue. Hind edge blue. Contact between eyes short (Fig 211c). From the side, frons pale blue. Eyes bluish slate-grey above and pale brown below (Fig 212c).
Thorax From above, dark brown with antehumeral stripes short and pale blue (Fig 211c). From the side, dark brown with two narrow, distinctly angulated, pale-blue bars. Legs black (Fig 212c).
Abdomen From above, tightly constricted at segment 3 creating a 'waisted' appearance (Fig 211c). Black with two pairs of blue spots on each segment, those near the tip being rectangular. Segments 1 & 2 predominantly blue. From the side, the whole of segment 3 is blue (Fig 212c), the remainder being black with conspicuous blue spots.

Azure Hawker

Wings Clear with costa faintly yellow. Pterostigmata long, narrow and dark greyish-brown.

Female 'blue' form (Fig 205)
Head Similar to the male but with eyes pale pinkish-brown (Fig 211d).
Thorax Similar to the male but with antehumeral stripes yellowish, very fine and often absent (Fig 211d).
Abdomen From above, similar to the male but base of the abdomen not constricted (Fig 211d). All of the blue markings are smaller in size, creating a generally darker appearance. The large, blue areas of segments 1 & 2 in the male are absent. Segment 2 with bold, dagger-like, vertical marking (Fig 107d). From the side, similar to the male but blue markings paler (Fig 211d).
Note: In cool or dull conditions the blue markings in both sexes can be reduced in intensity.
Wings Similar to the male.

Female 'yellow' form (Fig 215)
Similar to the female 'blue' form but with all blue markings replaced with dull yellow.

Similar species
Its very restricted geographical range means that there are only two similar species which could be found in company with the Azure Hawker.

Male and female 'blue' form
See Dragonfly Group 3.
Common Hawker male and female 'blue' form (p **175**, 177).
Southern Hawker male blue-spotted immature phase (p **175**, 179).

Fig. 138 Typical habitat of the Azure Hawker. Bridge of Grudie, West Ross

Azure Hawker

Female 'yellow' form
See Dragonfly Group 4.
Common Hawker female 'yellow/green' form (p **173**, **184**).
Southern Hawker female and adult male (p **183**, **186**).

Behaviour/field tips
This is a notoriously difficult species to find and rightly has earned the title of the Dragonfly-hunter's 'Holy Grail'. The terrain on which it breeds is boggy and extremely difficult (potentially dangerous) to negotiate. Add this to the dramatically changeable weather in the Highland valleys, where sunshine is often at a premium, and one has a worthy adversary indeed. On several occasions I have driven to the Highlands from Norfolk, sat in the car for three days watching the rain fall and then driven home empty-handed. So be prepared for failure!

Females may be found by diligent searching as they lay their eggs into mosses and other soft substrates around the margins of bog pools, but success is rare. The males can more easily be found in these situations as they fly swiftly from one pool to the next in search of females, but they land only infrequently. Perhaps the greatest chance to see the insects at close quarters is to wait until the sun becomes obscured by cloud. One should then search the surfaces of large rocks, dead trees and fence-posts as, in such circumstances, both sexes may be found basking openly in preparation for the re-emergence of the sun. If present, stands of trees or wooded tracks provide sheltered areas in which the adults often hunt. It is inadvisable to search the Azure Hawker's territory alone. The terrain can be treacherous, and it is very easy to stumble or fall partially into one of the concealed deep holes that are frequent in such places. One should always carry a mobile 'phone.

Southern Hawker
Aeshna cyanea (Müller, 1764)

Flight period
Usually from mid-June to mid-October. Optimum from early July to early September.

Distribution and status
England and Wales: Widespread and often common throughout except for the higher ground of the Pennines, the Lake District and Snowdonia. It is present in the island of Anglesey but is probably absent as a breeding species from the Isle of Man. It has not been recorded from the Isles of Scilly.
Scotland: It has a strong foothold centred along a line approximately between Kintyre and Argyll Main in the south-west and East Ross and Moray in the north-east. There are scattered records from elsewhere and its range is stated to

Southern Hawker

Fig. 139 Southern Hawker. Thompson Common, West Norfolk

be expanding in Scotland by Cham *et al.* (2014). It is absent from all of the off-shore islands.
Ireland: Absent.

Habitat preferences
This species usually breeds in wooded situations that contain shallow, and often shaded, pools. It is often one of the first dragonflies to colonise new ponds sited in gardens and parks etc. where they are at least partly surrounded by trees. Cham *et al.* (2014) also state that, in Scotland at least, it can inhabit small coastal ponds along streams where freshwater seepage dilutes their salinity.

Description
Length *c.* 65mm; wingspan *c.* 90mm.

Adult male (Fig 217)
Head From above, frons yellow with a black, T-shaped, central marking (Fig 220e). Eyes blue with broad contact. From the side, eyes blue above and green below. Frons yellow (Fig 220e).
Thorax From above, black, or very dark brown, with antehumeral stripes broad, bold and greenish-yellow (Fig 220e). From the side, very dark brown with a marbled pattern of pale-green bars and spots (Fig 221e). Legs dark brown or black.
Abdomen From above, ground colour black or very dark brown. Segment 2 with a conspicuous, downward-pointing, greenish-yellow triangle. Segment 3 is constricted to form a 'waisted' appearance (Fig 220e). Segments 2–7 each

Southern Hawker

with a pair of large, pale-green spots below pairs of small, pale-green triangles. Segment 8 has a pair of conjoined bright blue spots and segments 9 & 10 each have a bright blue bar (Fig 217). From the side, dark brown with a series of blue spots below a row of pale-green spots from segments 2–7. On segments 8–10 all markings are blue rather than green (Fig 217).
Wings Clear with pterostigmata short and black (Fig 217).

Immature male (Fig 210)
Similar to adult but all abdominal spots pale lilac in colour.

Female (Fig 218)
Similar to the male but lacks the constricted 'waist' and all abdominal spots are bright green. Also, eyes brown rather than blue and pterostigmata pale brown rather than black.

Similar species
Adult male
See Dragonfly Group 4.
The blue markings at the tip of the abdomen should preclude confusion with other green-spotted species.

Immature male
See Dragonfly Group 3.
Hairy Dragonfly male (p 174, **179**).
Emperor Dragonfly male (p 175, **179**).
Azure Hawker adult male and female 'blue' form (p 175, **179**).
Common Hawker adult male and female 'blue' form (p 176, **179**).
Migrant Hawker male (p 178, **179**).

Female
See Dragonfly Group 4.
Hairy Dragonfly female (p 183, **186**).

Fig. 140 Larvae sometimes abound in such places. A pingo at Thompson Common, West Norfolk

Southern Hawker

Emperor Dragonfly female (p 183, **186**).
Azure Hawker female 'yellow' form
(p 183, **186**).
Common Hawker female 'yellow/green'
form (p 184, **186**).
Migrant Hawker female (p **186**).

Behaviour/field tips
Individuals of this species are very much 'loners' and are aggressive towards other large dragonflies, including those of their own species, if encountered in their feeding haunts or breeding sites. When a male is observed hunting along a gloomy woodland ride, or a female is watched ovipositing into a moss-covered dead log in the margin of a shaded woodland pond, it is easy for one's imagination to carry oneself back to a primitive forest in the Carboniferous period where their ancestors performed the same rituals millions of years ago. The highly inquisitive males, who often inspect humans 'eye-to-eye', appear to resent one's presence in their home whilst asking the question 'What are you?' Such meetings should be savoured.

Both sexes can be found basking openly during cool conditions and the females often betray their presence by the audible rustle of their wings whilst laying eggs. The regular bouts of hovering during flight offer spectacular photo-opportunities.

I often have found males of this species in light-trap catches from the Rothamsted Insect Survey, so suggesting the species can continue to hunt into the hours of darkness.

Brown Hawker
Aeshna grandis (Linnaeus, 1758)

Flight period
Usually early June to late September. Optimum from early July to mid-August.

Distribution and status
England: Widespread and fairly common south of a line approximately between West Lancashire in the west and North-east Yorkshire in the east. North of this, the species is very scarce or absent. It is rare in the Isle of Man. To the south it appears to be more-or-less absent from Devon and Cornwall but is found commonly from lowland areas elsewhere. It has not been reported from the Isles of Scilly.
Wales: Found mainly in the central and eastern counties.
Ireland: Recorded only sparsely from northern and south-western counties but elsewhere it is widespread and fairly common.
Scotland: A single record from West Lothian in 2006 (Cham *et al.* 2014).

Habitat preferences
Still-waters, canals and very slow-moving rivers in lowland areas. It appears to be absent from high altitudes.

Brown Hawker

Fig. 141 Brown Hawker beautifully hidden against the trunk of a lofty pine. Holt Country Park, West Norfolk

Description

Length *c*. 70mm; wingspan *c*. 90mm.
Male (Fig 194)
Head From above, frons dull yellow with a brown, T-shaped, central marking. Eyes blue (Fig 194). From the side, frons pale brown (Fig 92). Eyes with broad contact. Blue above and pale brown below.
Thorax From above, brown with small, blue patches at the base of each wing (Fig 198a). Antehumeral stripes absent. From the side, there is a conspicuous cream or lemon-yellow stripe on each of segments 2 & 3. Legs dark brown (Fig 199a).
Abdomen From above, mainly brown and constricted at segment 3 giving a 'waisted' appearance (Fig 194). At the base, segment 2 with a horizontal pair of blue spots (Fig 194). Other segments each with a horizontal pair of clear, fine, cream lines. Towards the tip, these become larger and more triangular with the edges less well defined. From the side, brown with a series of blue spots along the lower half.
Wings Strongly tinged orange-brown with pterostigmata dark amber (Fig 194).

Female (Fig 195)
Head Similar to male but with eyes pale brown (Figs 198a; 199a).
Thorax Similar to male.

Brown Hawker

Abdomen Similar to male but not 'waisted'. Blue spots on segment 2 replaced by cream (Figs 198a; 199a).
Wings Similar to male.

Similar species
See Dragonfly Group 1.
Norfolk Hawker (both sexes) (p **172**, 173).

Behaviour/field tips
This large, elegant hawker is most often seen in flight as it hunts along sheltered hedgerows and woodland edges, rides and clearings. When so doing, the distinctive brown wings make it easy to identify. Once on the wing it will often continue to feed through dull periods when the sun is obscured by cloud. They rarely perch in the open and prefer dense vegetation close to the ground or the branches of tall trees. Here they can be very difficult to find (Fig 141). They are extremely alert and will usually take off at the slightest disturbance. They appear to be aware of one's presence long before one is aware of theirs. Probably the best way to photograph the species is to follow an individual until it decides to rest, making as precise as possible a note of where it landed and approach the spot very carefully. Doing so with one's back to the sun is advisable as the dragonfly will almost inevitably have its open wings facing the observer. Always be very aware not to cast a shadow onto, *or near,* the insect as this will immediately put it to flight. Many attempts may have to be made before success is achieved.

Common Hawker
Aeshna juncea (Linnaeus, 1758)

Flight period
Usually from early June to early October. Optimum from early July to mid-September.

Distribution and status
England: Widespread and often common in suitable habitats west of a line approximately between East Sussex in the south and South-east Yorkshire in the north. Elsewhere it is either extremely localised or absent, though it is recorded regularly from a few sites in East Anglia. Locally common in the Isle of Man. Absent from the Isles of Scilly.
Scotland: Common throughout, including the larger off-shore islands, though it is scarce in the Shetland Islands.
Wales: Widespread and usually common throughout in suitable localities, including those in the island of Anglesey.
Ireland: Localised but common in suitable habitats.

Habitat preferences
As its alternative name of Moorland Hawker suggests, this is truly a denizen of moorland and heathland containing bog pools where the females lay their eggs. It has also been recorded as

Common Hawker

breeding in larger lakes and even garden ponds where these are in, or adjacent to, its usual habitats. Sheltered woodland provides suitable hunting areas.

Description
Length *c*. 65mm; wingspan *c*. 90mm.
Male (Fig 206)
Head From above, frons yellow with a black, T-shaped, central marking. Eyes slate-blue with a broad point of contact (Fig 211e). From the side, frons yellow. Eyes slate-blue above and brown below (Fig 101e).
Thorax From above, dark brown with the antehumeral stripes long, narrow and pale blue or bluish-cream in colour. These are slightly divergent, being closest near the thorax and separating towards the head (Fig 107e). From the side, dark brown with two straight, blue bars (Fig 212e). Legs dark brown or black.
Abdomen From above, black or very dark brown and tightly constricted at segment 3, creating a 'waisted' appearance (Fig 107e). Segments 1 & 2 each with a broad, blue, horizontal bar (Fig 211e). Remaining segments each with a pair of blue spots below a pair of small, pale-yellow triangles (Fig 206). From the side, segment 3 has the upper half black or very dark brown and the lower half blue (Fig 212e). Subsequent segments black or very dark brown with a series of blue spots. The lower halves of the yellow triangles on the dorsal surface are also visible.
Wings Clear with costa usually strongly edged yellow (Fig 212e). Pterostigmata long, narrow and pale brown. (Fig 99).

Female 'blue' form (Fig 207)
Superficially very similar to male but abdomen not 'waisted' in appearance. Also the antehumeral stripes are absent, or almost so (Fig 211f).

Female 'yellow/green' form (Fig 216)
Similar to female 'blue' form but all blue markings replaced with yellowish-green (Figs 220d; 221d).

Similar species
Male and female 'blue' form
See Dragonfly Group 3.
Hairy Dragonfly adult male (p 174, **176**).
Emperor Dragonfly adult male (p 175, **176**).
Azure Hawker male and female 'blue' form (p 175, **176**).
Migrant Hawker male (p **176**, 178).
Southern Hawker immature male (p **176**, 179).

Female 'yellow/green' form
See Dragonfly Group 4.
Hairy Dragonfly female (p 183, **184**).
Emperor Dragonfly female (p 183, **184**).
Azure Hawker female 'yellow' form (p 183, **184**).
Southern Hawker adult male and female (p **184**, 186).
Migrant Hawker female (p **184**, 186).

Behaviour/field tips
The Common Hawker is a fast, powerful and acrobatic flier. It is capable of terrific turns of speed and can change direction in 'the blink of an eye'. This, coupled with its rather dark appearance, makes it a difficult species to follow in flight. The

Common Hawker

Fig. 142 Common Hawker. Longmynd, Shropshire

Fig. 143 Ideal hunting grounds for Common Hawker. Bridge of Grudie, West Ross

Common Hawker

males are seemingly tireless in their aerial pursuit of prey and/or females; settling only infrequently. It flies even in quite dull conditions and, unlike most other hawkers, continues to be active whilst the sun is obscured by cloud. The females can be found ovipositing by scanning carefully along mats of floating vegetation using binoculars. Ripples on the water or, at close quarters, the sound of rustling wings will often indicate their presence.

Migrant Hawker
Aeshna mixta Latreille, 1805

Flight period
Usually from late July to late October. Optimum from early August to late September.

Distribution and status
England and Wales: Widespread and often common, though less so on higher ground. It is recorded regularly from the Isle of Man and Anglesey and occasionally from the Isles of Scilly.
Scotland: Absent from much of the country, though recent records suggest an expansion northward into some southern

Fig. 144 Migrant Hawker. Wat Tyler Country Park, South Essex

Migrant Hawker

Fig. 145 Migrant Hawker abounds along such rides in late Summer and early Autumn. Holkham Meals, West Norfolk

counties (Cham *et al.*, 2014).
Ireland: Restricted mainly to the counties of Clare and Limerick in the west and the coastal hinterlands of the east and south. There is a scattering of records from elsewhere in the Republic.

Habitat preferences
Breeds in a wide range of habitats, particularly in still or slow-moving waterways with abundant marginal vegetation. It is also tolerant of brackish situations near the coast. The rides of nearby woodlands are used extensively for hunting.

Description
Length *c*. 65mm; wingspan *c*. 85mm.
Adult male (Fig 208)
Head From above, contact between eyes broad. Frons yellow with a large, black, T-shaped, central marking. Eyes intense blue (Fig 211g). From the side, frons yellow. Eyes blue above and pale brown below (Fig 212g).
Thorax From above, dark brown with antehumeral stripes short, broad and pale yellow (Fig 211g). From the side, dark brown with two broad, yellow bars tinged blue towards the wings. Between these is a series of small, yellow spots.

Migrant Hawker

Legs dark brown or black (Fig 212g).
Abdomen From above, very dark brown or black. At the base, segment 2 with a conspicuous yellow, 'golf tee'-shaped marking above a broad, bright-blue, horizontal band. Remaining segments each have a pair of small, yellow, triangular markings above a pair of bright-blue spots (Fig 211g). From the side, segment 3 mainly blue (Fig 212g). Subsequent segments dark brown or black with a series of bright-blue spots. Superior claspers disproportionately long (Fig 208).
Wings Clear with pterostigmata orange-brown.

Immature male (Fig 209)
Similar to the adult but blue markings replaced with lilac. Eyes are pale brownish-grey rather than blue.

Female (Fig 219)
Markings similar to male but blue replaced by pale yellow; eyes pale brown; base of abdomen with a pair of large, yellow spots replacing blue bar on segment 2 (Fig 220f).

Similar species
Adult and immature male
See Dragonfly Group 3.
Hairy Dragonfly adult male (p 174, **178**).
Emperor Dragonfly adult male (p 175, **178**).
Azure Hawker adult male and female 'blue' form (p 175, **178**).
Common Hawker adult male and female 'blue' form (p 176, **178**).
Southern Hawker immature male (p **178**, 179).

Female
See Dragonfly Group 4.
Hairy Dragonfly female (p 183, **186**).
Emperor Dragonfly female (p 183, **186**).
Azure Hawker female 'yellow' form (p 183, **186**).
Common Hawker female 'yellow/green' form (p 184, **186**).

Behaviour/field tips
A large gathering of dragonflies hunting together in woodland clearings or along rides and hedgerows during late summer is sure to be of this species. Their flight is deliberate, purposeful and punctuated with frequent bouts of hovering. They settle regularly on vegetation, both high and low, and can then often be approached with confidence. During the early morning, especially as summer gives way to autumn, large numbers can sometimes be found basking together on surfaces such as fences or fallen dead tree-trunks that face the rising sun. In such circumstances, when the insects are cool, the blue markings of the male may be less intense and more like those of an immature individual. When ovipositing, the females are rather secretive and often difficult to find. However, at this time, males can often be seen hovering low over bankside vegetation. Watch a male closely and one may be led to his suitor.

Norfolk Hawker
Anaciaeshna isosceles (Müller, 1767)

Fig. 146 Norfolk Hawker
Strumpshaw Fen, East Norfolk

Habitat preferences
The species usually breeds in ditches that traverse fens and marshes. It requires very clean water with abundant floating and/or emergent vegetation. The presence of Water-soldier (*Stratiotes aloides*) is an indicator of such water quality and the Norfolk Hawker is therefore often associated with this plant. However, it appears that the absence of Water-soldier does not necessarily preclude the presence of the dragonfly. It has been occupying pools and small lakes where Water-soldier is present, such as the mature, disused quarry at Paxton Pits.

Flight period
Usually from late May to late July. Optimum from mid-June to mid-July.

Distribution and status
England: The Norfolk Hawker is a very localized and scarce species (Red Data Book Category 1; Wildlife and Countryside Act (1981) Schedule 5). As such it is illegal to catch or handle the insect without an official permit from English Nature.

Traditionally, its distribution was restricted to a few of the Broads in east Norfolk and east Suffolk. However, in recent years there have been regular sightings in Kent (Westbere Lakes) and Cambridgeshire (Paxton Pits). Here, breeding has been proven by the discovery of exuviae. Single specimens from elsewhere (e.g. Lincolnshire and Bedfordshire) probably represent migrating individuals from mainland Europe.

Scotland, Wales and Ireland: Absent.

Fig. 147 A typical ditch in which Norfolk Hawker may breed. Strumpshaw Fen, East Norfolk

Norfolk Hawker

Description
Length *c*. 65mm; wingspan *c*. 90mm.
Male (Fig 196)
Head From above, contact between eyes broad. Frons pale brown. Eyes bright metallic green (Fig 196). From the side, Frons pale brown. Eyes metallic green above and pale green below.
Thorax From above, brown and without antehumeral stripes (Fig 198b). From the side, brown with two broad, greenish-yellow panels. Legs dark brown or black.
Abdomen From above, brown. At the base there is a conspicuous cream, vertical marking in the shape of an isosceles triangle on segment 2 (Fig 198b). Segments 3–7 each with a fine, black, central line and black interseg-mental divisions. Segments 8 & 9 with central line expanded to form elongated black spots. Segment 10 unmarked (Fig 196).
Wings Clear but with a small amber patch at base of hindwings. Pterostigmata dark amber (Fig 196).

Female (Fig 197; Fig 146)
Very similar to male but abdomen with cream triangle on segment 2 edged with black (Fig 197). All black abdominal markings more conspicuous. Amber basal patch of hindwing usually absent.

Similar species
Both sexes
See Dragonfly Group 1.
Brown Hawker both sexes (p 172, **173**).

Behaviour/field tips
Like all of the hawkers, this is a powerful and elegant flier. During its hunting forays it often hovers whilst holding its body perfectly still, thus offering good views of the remarkable metallic green eyes. It appears to be quite inquisitive; sometimes gazing straight at the observer. As a result of this behaviour, spectacular 'flight-shot' photographs may be available.
Perhaps the best time to examine and photograph perched individuals is during sunny weather with intermittent cloud. Soon after the sun becomes obscured, they land openly on tall vegetation or bushes without attempting to conceal themselves and can then be approached with confidence.
When hunting, adults may range quite widely across sheltered scrubland, hedges and woodland edges and clearings close to their breeding grounds. Tarmac-covered lanes and tree-fringed car parks often provide suitable alternatives to these more natural habitats. Here the warmth from the surface substrate and the shelter of surrounding vegetation provide excellent areas in which to hunt for small insects.

Emperor Dragonfly
Anax imperator Leach, 1815

Flight period
Usually from late May to late August. Optimum from mid-June to late July.

Distribution and status
England: With the exception of the upland areas of the Pennines and the Lake District, it is widespread and sometimes common in suitable districts south of a line approximately between Cumberland in the west and South Northumberland in the east. Resident but localised in the Isle of Man. It has been recorded recently in the Isles of Scilly.
Scotland: Distributed very locally along the south-west coastal hinterlands of Dumfriesshire, Kirkcudbrightshire and Wigtownshire. There are scattered records from elsewhere, including one from as far north as Argyll Main (Cham et al., 2014). It is absent from all of the offshore islands.
Wales: Widespread and fairly common throughout.
Ireland: First time as recently as 2000 (Thompson & Nelson, 2014), since when there have been widely scattered records from across the south-eastern half of both Northern Ireland and the Irish Republic.

Habitat preferences
The Emperor Dragonfly is a great pioneer species that is often the first to

Fig. 148 Emperor Dragonfly. New Forest, South Hampshire

Emperor Dragonfly

utilise newly flooded pits, quarries and even garden ponds. More typically, it inhabits well-vegetated lakes, pools, large ditches, canals and very slow-moving rivers. An abundance of floating vegetation is preferred as the females usually perch on the leaves of aquatic plants or weed-mats whilst egg-laying (Fig 148). It has a wide tolerance of water types and can even be found breeding in brackish pools near the coast.

Description
Length *c*. 75mm; wingspan *c*. 105mm.
Adult male (Fig 203)
Head From above, eyes with contact broad. Frons pale blue with a small, black, triangular, central mark near the eyes. Eyes bright blue (Fig 211b). From the side, upper part of frons pale blue (Fig 212b).
Thorax From above, apple-green with a pair of conspicuous blue, horizontal, 'back-to-back', triangular markings near the wings (Fig 211b). From the side, uniform apple-green (Fig 212b). Legs dark brown or black.
Abdomen From above, segment 1 green but the remainder is mainly blue with a black, central band and black, segmental divisions (Fig 203). From the side, mainly blue above with a series of extensive black stripes along lower half.
Wings Clear with costa yellow (Fig 211b) and pterostigmata slender, long and amber-brown.

Immature male
As above but all blue replaced with yellowish-green; thus resembling a female. This colouration is also expressed in adult males that have recently been subjected to low temperatures or darkness.

Female (Fig 214)
Markings similar to male but all blue colouration replaced with apple-green and central abdominal stripe broader.

Similar species
Adult male
See Dragonfly Group 3.
Hairy Dragonfly adult male (p 174, **175**).
Azure Hawker male and female 'blue' form (p **175**).
Common Hawker male and female 'blue' form (p **175**, 176).
Migrant Hawker male (p **175**, 178).
Southern Hawker immature male (p **175**, 179).

Female and immature male
See Dragonfly Group 4.
Hairy Dragonfly female (p **183**).
Azure Hawker female 'yellow' form (p **183**).
Common Hawker female 'yellow/green' form (p **183**, 184).
Southern Hawker both sexes (p **183**, 186).
Migrant Hawker female (p **183**, 186).

Behaviour/field tips
The Emperor Dragonfly is simply a master of the air. It follows a regular beat around its territory with controlled determination and is capable of both lightening-fast attacks and gracefully poised hovers. Its aggression is second to none and it will tackle prey as large as

Emperor Dragonfly

Fig. 149 Emperor Dragonfly. A hunting male. New Forest, South Hampshire

darter dragonflies and even an unwary Four-spotted Chaser. On finding an individual, simply enjoy this embodiment of 'all things dragonfly'. Its size and behaviour make it easy to spot and wonderful photo-opportunities will present themselves during its frequent bouts of hovering. Also note that in flight the abdomen is often held in a downwardly curved position. It can sometimes be found at rest on, or amongst, bushes but one's approach must be extremely stealthy to avoid disturbing what is a very wary creature.

Hairy Dragonfly
Brachytron pratense (Müller, 1764)

Flight period
Usually late April to early July. Optimum from mid-May to mid-June. This is the first of the hawker dragonflies to emerge each year.

Distribution and status
England: Widespread and locally common east of a line approximately between Dorset in the south and Southeast Yorkshire in the north. There are also colonies based in and around South and North Somerset and Cheshire. It is absent from the Isle of Man and the Isles of Scilly.
Scotland: Absent from most of the country and her off-shore islands. However, there are isolated populations centred on the counties of Dumfriesshire, Kirkcudbrightshire and Wigtownshire and the extreme west of Kintyre,

Fig. 150 Hairy Dragonfly
Upton Fen, East Norfolk

Argyll Main and West Inverness-shire.
Wales: Restricted mainly to the coastal hinterlands of the southern counties and

Hairy Dragonfly

a few of those in the north-west, including the island of Anglesey.
Ireland: Present in scattered localities throughout.
It should be stressed here that this once quite scarce species appears to be undergoing a rapid expansion in its known range in the British Isles (Cham *et al.*, 2014).

Habitat preferences

High quality, unpolluted, still or very slow-flowing waters with a rich floral diversity are prerequisite for this species. Open waters with dense and varied, marginal emergent vegetation is usually preferred. These come in the form of ponds, lakes, canals and drainage ditches such as those found in fens and grazing marshes. Ideally, adjacent woodland and scrub provide areas in which the adults can shelter and hunt.

Description

Length *c*. 55mm; wingspan *c*. 75mm.
Adult male (Fig 202)
Head From above, contact between eyes short. Frons pale yellow with a black, T-shaped, central marking. Eyes blue (Fig 211a). From the side, frons yellow. Eyes blue (Fig 212a).
Thorax From above, dark brown and conspicuously downy with antehumeral stripes broad and pale yellowish-green (Fig 211a). From the side, downy with three bold, apple-green panels, the central one of which forms a slender rectangle. Legs dark brown or black (Fig 212a).

Fig. 151 Typical fenland habitat for Hairy Dragonfly. Sculthorpe Moor, West Norfolk

Hairy Dragonfly

Abdomen From above, black. Each segment with a pair of blue spots below a pair of yellow, slender, horizontal lines. Superior claspers disproportionately long (Fig 202). From the side there is a linear series of blue spots along the lower half. The blue spots from the dorsal surface are also visible.
Wings Clear with pterostigmata long, slender and dark amber (Fig 202).

Female (Fig 213)
Head Blue colour of eyes less extensive than in the male and replaced mainly by greenish-brown (Figs 220a; 221a).
Thorax From above, chestnut-brown with antehumeral stripes pale yellow and very short (Fig 220a). From the side, similar to male (Fig 221a).
Abdomen All blue markings of the male replaced by pale yellow (Fig 213).
Wings Heavily tinged with dark amber at bases and along costa (Figs 213; 220a).

Similar species
Male
See *Dragonfly Group* 3.
Emperor Dragonfly adult male (p **174**, 175).
Azure Hawker male and female 'blue' form (p **174**, 175).
Common Hawker male and female 'blue' form (p **174**, 176).
Migrant Hawker male (p **174**, 178).
Southern Hawker immature male (p **174**, 179).

Female
See *Dragonfly Group* 4.
Emperor Dragonfly female and immature male (p **183**).
Azure Hawker female 'yellow' form (p **183**).
Common Hawker female 'yellow/green' form (p **183**, 184).
Southern Hawker female and adult male (p **183**, 186).
Migrant Hawker female (p **183**, 186).

Behaviour/field tips
The males of this species are most often seen patrolling the edges of ponds and ditches as they search for females. This they do with purposeful, though seemingly erratic, low flights closely adjacent to, or amongst, bankside vegetation. They are difficult to follow and seldom hover or land. The female is only rarely encountered in such situations but can sometimes be found ovipositing – the tell-tale sign of her presence being the gentle sound of rustling wings. Females can also often be seen settled conspicuously on tall waterside vegetation. However, much of their time is spent away from the breeding sites, often with males, hunting in nearby shrubby areas and wooded tracks and clearings. Here they can often be seen perched openly on the branches and leaves of tall bushes or trees. Perhaps the best way to photograph them is to watch closely as they hunt until the sun is obscured by cloud. They will then land almost immediately. If the insect is disturbed by a clumsy approach it will usually take only a short flight to an alternative resting place and a second opportunity for success should arise.

Club-tailed Dragonfly
Gomphus vulgatissimus (Linnaeus, 1758)

Flight period
Usually from early May to early July, though this may be later in the west of its range. Optimum from mid-May to late June.

Distribution and status
England: This generally uncommon riverine species is restricted to suitable areas along the rivers Severn, Thames, Wye and Arun, and some of their major tributaries. Absent from the Isle of Man and the Isles of Scilly.
Wales: Very restricted in its distribution and found only along the catchment areas of the rivers Dee, Teifi, Twyi and Wye. Absent from the island of Anglesey.
Scotland and Ireland: Absent.

Habitat preferences
The Club-tailed Dragonfly requires unpolluted, slow- to medium-flowing rivers along which there are areas such as eddies, backwaters and meanders that cause the deposition of a layer of silt in which the larvae can burrow and hunt. A good cover of bankside and adjacent trees is essential for hunting, shelter and pairing sites for the adults. This is especially so for the females that use such

Fig. 152 Club-tailed Dragonfly, freshly emerged male. Goring, Oxfordshire

Club-tailed Dragonfly

places for resting between bouts of egg-laying.

Description
Length *c*. 65mm; wingspan *c*. 65mm.
Adult male (Fig 222)
Head From above, there is no contact between the eyes. Frons bright yellow. Eyes pale green joined by a bright, yellow-green bar (Fig 224b). From the side, frons bright yellow, heavily edged with black. Eyes entirely pale green (Fig 224c).
Thorax From above, black with antehumeral stripes broad and pale green. Between antehumeral stripes there is a bold, pale-green, arrowhead-shaped marking (Fig 224b). From the side, mainly pale green with segments 2 & 3 separated by a fine, black line (Figs 224a; 224c). Legs black.
Abdomen Distinctly club-shaped. From above, black with pale-green markings comprising a horizontal bar on segment 1; a large, thistle-shaped marking on segment 2; a prominent central stripe on segments 3–5 and an elongated, central triangle on segments 6 & 7. Segments 8 & 9 each with conspicuous yellow inter-segmental divisions and a pair of lateral

Fig. 153 River Thames at Goring, Oxfordshire

Club-tailed Dragonfly

yellow spots (Fig 222). Segment 2 with a pair of yellow, laterally protrusive spherical structures known as 'genital lobes' (Figs 222; 224b), the prominence of which are diagnostic of this species. From the side, segments 1 & 2 black above and yellow below; segments 3–6 black with yellow intersegmental divisions; segments 7–9 black, each with a large, yellow spot (Fig 224a). Beneath segment 2, the protruding secondary genitalia are clearly visible as a series of black ridges (Figs 224a; 224c).
Wings Clear with costa yellow (Fig 224b) and pterostigmata dark brown. Inner margin of hindwing acutely indented (Fig 224b).

Immature male (Fig 152)
As above but with all green markings replaced by yellow.

Female (Fig 223)
Similar in general appearance to male but all green markings replaced with yellow, abdomen without protrusive secondary genitalia and genital lobes, and inner margin of hindwing not indented.

Similar species
None. The separation of the eyes (similar to those of a damselfly) is a unique feature amongst the British dragonflies. See Dragonfly Group 5.

Behaviour/field tips
Even in areas where populations are strong this can sometimes be an infuriatingly difficult species to find as an adult. The larvae have a highly synchronised emergence from the water and they, or their exuviae, can often be found in good numbers clinging to riverside plants or structures at this time. However, the adults disperse quickly into woods and hedgerows, sometimes long distances from their breeding localities. The lucky observer may then find them hunting in woodland clearings or orchards, but disturbance will see them flee immediately to the canopy and out of sight. On reaching maturity, both sexes return to their riverside stations. From here they may suddenly appear, as if from nowhere, and the males may be seen flying slowly and close to the surface of the water in search of prey or potential mates. The forays of the females to the water are usually very brief and their visits appear to be solely for the purpose of laying eggs. If disturbed, both sexes usually beat a hasty retreat to the neighbouring treetops. On days of intermittent sunshine, males may sometimes be disturbed from bankside vegetation; their fluttering flight being very distinctive. If one is very fortunate, they may be found sunbathing on shrubs, fenceposts or patches of bare earth.

Golden-ringed Dragonfly
Cordulegaster boltonii (Donovan, 1807)

Fig. 154 Golden-ringed Dragonfly. New Forest, South Hampshire

Flight period
Usually from late May to late September. Optimum from mid-June to early August.

Distribution and status
England: Common on the heaths and moors of the south-west and the south and in similar habitats across the north of the country. It has not been recorded in the Isle of Man or the Isles of Scilly.
Scotland: Widespread and locally common but absent from much of the Central Lowlands and low-lying areas of the east of the country. There are outlying populations in the Orkney Islands but absent from the Outer Hebrides and the Shetland Islands.
Wales: Found commonly throughout all of the upland areas including those in the island of Anglesey.
Ireland: Absent.

Habitat preferences
This is very much a species of fast-flowing rivulets and small rivers that traverse heathland, moorland with light woodland and scrub. It also sometimes breeds along drainage ditches and mountain flushes.

Description
Length *c.* 75mm; wingspan *c.* 90mm.
Both sexes (Figs 200; 201)
Head Eyes bright metallic green with short contact. Frons yellow (Fig 154).
Thorax and abdomen
Black with spectacular bright golden markings. Abdomen of male distinctly club-shaped (Fig 200).
Wings Clear with pterostigmata long, narrow and black (Fig 203).

Similar species
None. The Golden-ringed Dragonfly is unlike any other species found in the British Isles.

Fig. 155 Golden-ringed Dragonfly ovipositing female. Bridge of Grudie, West Ross

Golden-ringed Dragonfly

See Dragonfly Group 2 (p 174).

Behaviour/field tips
Territorial males spend most of their time patrolling stretches of streams in open, sunny areas. Whilst doing so, they fly very close to the water's surface and perch frequently and openly on bankside vegetation. They can then be examined and photographed with ease so long as they are not approached too closely as they are very alert and will quickly take flight if alarmed.

The females are more secretive, especially when laying eggs. This is done in both sunny and slightly shaded areas and they should be sought by walking quietly along the streams looking for any unusual movement. They oviposit by hovering close to the water with their abdomen held pointing straight down and partially submerged. The long, pointed ovipositor is then repeatedly 'stabbed' downwards into the substrate where the eggs are deposited. This action is reminiscent of the needle on a sewing machine and is fascinating to watch. It can also be photographed provided the ambient light levels allow the use of high shutter speeds (Fig 155). Hunting adults can often be found some distance from the breeding sites and can sometimes be seen patrolling along woodland edges and their more open tracks.

Downy Emerald Dragonfly
Cordulia aenea (Linnaeus, 1758)

Flight period
Usually from early May until the end of July. Optimum from late May to early July.

Distribution and status
England: The Downy Emerald is widespread in the south-eastern counties approximately as far north as Buckinghamshire and Northamptonshire and west as far as Dorset and Gloucestershire. Further west, there are outlying colonies in South Somerset and Devon and further north there are a few populations in Shropshire, Staffordshire, Cheshire and Westmorland. Elsewhere, there is a single colony at a private site in East Norfolk. It has not been reported from

Fig. 156 Downy Emerald Dragonfly. Captain's Pond, West Norfolk

Downy Emerald Dragonfly

Fig. 157 A typical woodland breeding site. Bentley Woods, South Wiltshire

the Isle of Man.
Scotland: Restricted to a few localities in Argyll Main and East Inverness-shire. It is absent from the offshore islands.
Wales: Found only in the counties of Glamorgan, Merionethshire and Montgomeryshire. It is absent from the island of Anglesey.
Ireland: There are very isolated populations centred around the counties of South Kerry and in West Galway.

Habitat preferences

This very localised species is usually associated with nutrient-poor, deciduous woodland pools where the larvae can develop amongst a thick layer of decaying leaf litter. If allowed to mature, some flooded former mineral-extraction sites may become suitable provided there are fringes of trees. It may also breed in tree-lined canals. In Scotland, the Downy Emerald can be found at lochans and lochs within areas of ancient Caledonian pine forest and in Ireland it inhabits more exposed blanket bog pools adjacent to coniferous woodland. Wherever it breeds, nearby woodlands appear to be an essential requirement for shelter, hunting and maturation sites for immature individuals.

Description

Length *c*. 50mm; wingspan *c*. 65mm.
Male (Fig 225)
Head From above, frons green. Eyes with contact short. Brown, turning bright metallic green with maturity (Fig 231a). From the front, frons black without yellow markings (Fig 235a).
Thorax Dark metallic green, conspicuously downy and without diagnostic markings. Legs black (Figs 235a; 156).
Abdomen From above, dark metallic bronze-green and distinctly club-shaped (Fig 225). At the base, segment 2 bulbous and segment 3 constricted giving a 'waisted' appearance (Fig 225). There is a pair of tiny, yellow, horizontal lines at the junction of segments 2 & 3.

Downy Emerald Dragonfly

Superior claspers short and outwardly curved (Fig 233a). From the side, segment 2 yellow-orange below. Segment 3 with large, white ventral patch (Fig 232a).
Wings Clear but with bases tinged dark orange. Inner margin of hindwing acutely concave. Pterostigmata short and dark grey or black (Fig 225).

Female
Superficially similar to male but abdomen not club-shaped or 'waisted' at the base. Inner margin of the hindwing not concave (Fig 226).
From above, superior claspers short, thick and straight (Fig 233d). From the side, vulvar scale beneath segment 8 is usually not visible.

Similar species
Both sexes
See *Dragonfly Group 6*.
Northern Emerald Dragonfly (p **189**, 194).
Brilliant Emerald Dragonfly (p **189**, 192).

Behaviour/field tips
Its secretive nature, intolerance of dull weather, cryptic colouration and specific habitat requirements make this rather enigmatic dragonfly difficult to find and sometimes seemingly impossible to photograph. Much of its time is spent amongst the foliage of trees and tall bushes surrounding the breeding pools. Here it matures after emergence, hunts, rests and copulates. The females only return to the water to lay their eggs, and the males to patrol their marginal beats in search of prey or females. During these occasions they may both feed for brief periods on small flying insects before returning to the treetops. The males' patrol flights are usually very close to the surface of the water and are quite erratic, with much to-ing and fro-ing and short spells of hovering. They often fly so close to the water's surface that ripples are formed beneath the wings. The abdomen is usually held with the tip slightly above the horizontal, so giving a characteristic 'tail-up' view in profile. Males are very aggressive towards any other dragonflies of approximately their own size but tend to avoid larger species such as the Emperor Dragonfly. They are very inquisitive and often turn to face the human observer. In so doing, one is confronted by a pair of brilliant, metallic green eyes that appear 'other-worldly'. The memory of this wonderful sight remains with one for many a year. Females are far more secretive, spending most of their time searching for sunny marginal areas with submerged or short, emergent vegetation amongst which they lay their eggs – usually hovering whilst ovipositing. They are most visible during the early morning and late afternoon when the males are least active.
Perhaps the best chances to photograph the Downy Emerald occur whilst they hunt in woodland clearings and rides as they will then sometimes rest within reach of the camera. A cautious approach is essential as, if disturbed, the insect will fly rapidly to the canopy and the opportunity will be lost.

Northern Emerald Dragonfly
Somatochlora arctica (Zetterstedt, 1840)

Flight period
Usually from mid-May to late August. Optimum from mid-June to late July.

Distribution and status
England, Wales, Anglesey, the Isle of Man and the Isles of Scilly: Absent.
Scotland: It has a widespread but very localised distribution within an approximately triangular range between the counties of Dunbarton in the west to Moray in the east and then north-westwards to West Ross. There are a few scattered outlying colonies both to the south and north of this area. There has been a single record from the Island of Muck (Cham *et al.*, 2014) and a few recent records from the Isle of Mull. Apart from these, the species appears to be absent from the Scottish Isles.
Ireland: Restricted to the counties of South Kerry and West Cork.

Habitat preferences
The Northern Emerald is very much a creature of open moorland where it breeds in bog pools, flushes, seepages and ditches. It prefers sites where mosses are prevalent in which the adults lay their eggs and under which the larvae develop.

Fig. 158 Northern Emerald Dragonfly. Kinlochewe, West Ross

Northern Emerald Dragonfly

There may occasionally be very little open water. The nearby presence of trees and shrubs offers the adults places to mature, hunt and mate.

Description
Length *c*. 50mm; wingspan *c*. 65mm.
Male (Fig 229)
Head Eyes brown at first, later turning brilliant emerald-green with maturity. From the front, frons black with a prominent yellow spot on each side (Fig 235c).
Thorax Slightly downy. From above, dark metallic green tinged with blue (Fig 229). There is a fine, yellow or pale-brown, upward-pointing triangular marking emanating from the bases of the forewings and a yellow or cream spot near the head (Fig 231b). From the side, entirely dark metallic green. Legs black (Fig 232c).
Abdomen Dark metallic green tinged with blue (Fig 230). From above, segment 2 and part of segment 3 tightly constricted giving an elongated 'waisted' appearance. Junction of segments 2 & 3 with a pair of tiny, yellow or cream, horizontal lines. (Fig 229). Superior claspers characteristically 'pincer'-shaped

Fig. 159 Northern Emerald Dragonfly habitat at Glen Gour, West Inverness-shire

Northern Emerald Dragonfly

(Fig 233c).
Wings Clear but appearing bronze-tinged in certain lights. Pterostigmata brown. Hindwing with inner margin concave (Fig 229).

Female (Fig 230)
Head Similar to male.
Thorax Similar to male but yellow markings usually brighter (Fig 231b).
Abdomen From above, segment 2 with a pair of cream or yellow, horizontal lines. Junction of segments 2 & 3 with a pair of small, cream or yellow lines. Segment 3 with a pair of large, yellow spots (Fig 231b). Superior claspers fine and straight (Fig 233f). From the side, segment 2 with diagnostic, yellow or cream lines forming the outline shape of a Wellington boot. Segment 3 with the yellow spots from the upperside still visible (Fig 232c). Lower half is white, tinged with yellow on mature individuals. Vulvar scale, beneath segment 8 slightly protrusive (Fig 234a).
Wings Similar to male but inner margin of hindwing not concave (Fig 231b).

Similar species
Both sexes
See *Dragonfly Group 6*.
Downy Emerald Dragonfly (p 189, **194**).
Brilliant Emerald Dragonfly (p 192, **194**).

Behaviour/field tips
Because populations are usually quite small one should not expect to see many individuals on a single visit to a site. The inclement weather of the Scottish Highlands, and the often inhospitable terrain that the species inhabits, make this a challenging insect to find. Like the other emerald dragonflies, it is very sunshine dependent for its flight activities and will quickly seek the shelter of tree canopies during cloudy spells.

The males patrol the margins of breeding pools, in deliberate fashion, about a metre above the surface, often hovering for short periods. Occasionally they will swoop quickly upwards in a characteristic figure-of-eight pattern, either to pick off higher flying prey or to investigate an intrusive male. They are very intolerant of other dragonflies – even challenging the much larger Common Hawker. Such encounters usually end in submission and the Northern Emerald will leave to try his luck elsewhere. They are rarely seen flying over open water.

Hunting forays are usually at canopy level. When so engaged, the dragonfly can be observed at length through binoculars, but photography is extremely difficult as one has to compensate for the bright background of the sky whilst maintaining the required high shutter speed of the camera. However, in silhouette it can easily be identified as the thorax and large, bulbous first segment of the abdomen merge to create the impression of a huge thorax and a short 'tail'.

The females are less often seen than the males as they leave the sanctuary of the trees and return to the margins of the breeding ponds solely for the purpose of laying eggs.

Brilliant Emerald Dragonfly
Somatochlora metallica (Vander Linden, 1825)

Flight Period
Usually from late May to early August. Optimum from mid-June to mid-July. Populations in Scotland usually fly slightly later than those in southern England.

Distribution and status
It must be stressed that this insect is often secretive and very difficult to observe and is probably awaiting discovery at many sites within its known ranges.
England: Locally common at its strongholds centred on the counties of North Hampshire, Berkshire, Surrey and West and East Sussex. It is absent from the Isle of Man and the Isles of Scilly.
Scotland: There are populations at several sites centred on the county of Argyll Main and East-Inverness-shire. In both areas it can be found commonly in suitable localities. It has not been recorded from the offshore islands.
Wales and Ireland: Absent.

Habitat preferences
The Brilliant Emerald is usually associated with tree-fringed lakes, ponds, lochs and lochans, though it has also been found along tree-lined canals and the slower-moving sections of the rivers Arun and Ouse in south-east England. In some parts of Scotland it inhabits remote, sheltered hill lochans without surrounding trees (Cham *et al.*, 2014). Sheltered and shaded, muddy or mossy marginal areas are prerequisite for oviposition and larval development.

Description
Length *c*. 55 mm; wingspan *c*. 75mm.
Male (Fig 227)
Head Eyes brown at first, turning to bright metallic green with maturity (Figs 231c & d). From the front, frons with a conspicuous bright, yellow, elongated, 'U'-shaped marking which gives the appearance of a smiling face (Fig 235b).
Thorax Slightly downy and metallic emerald-green. From above, there is a fine, upward-pointing, triangular, yellow or pale-brown marking emanating from the bases of the forewings and a bright cream or yellow spot near the head (Figs 231c & d). From the side, bright metallic green without distinguishing markings. Legs black (Fig 232b).
Abdomen From above, constricted at segment 3 giving a 'waisted' appearance (Fig 227). Dark emerald-green with a pair of small, yellow lines either side of the junction between segments 2 & 3. Superior claspers long, straight and slightly inwardly pointing and each with a small, lateral spine. Inferior appendage long, triangular and clearly visible (Fig 233b).
Wings Clear with pterostigmata dark brown (Fig 227). Hindwings with inner margin concave (Fig 227).

Female (Fig 228)
Head and Thorax As in male.
Abdomen Mainly bright metallic green. From above, abdomen not 'waisted'. Segment 2 with a pair of tiny, yellow lines and a more conspicuous yellow or cream line at the junction of segments 2 & 3. Segment 3 with a pair of cream or

Brilliant Emerald Dragonfly

Fig. 160 Brilliant Emerald. Loch Bran, East Inverness-shire

yellow spots (Fig 231d). These may be absent in some individuals (Fig 231c). Superior claspers long and straight (Fig 233e). From the side, upper half of segment 2 with lines from dorsal surface still visible. Lower half with a series of three large, yellow patches. Segment 3 with lower half white, turning yellow with maturity. Vulvar scale below segment 8 long and protruding almost at right angles from the body (Fig 234b). *Wings* Hind margin not concave (Figs 231c & d). Clear with yellow suffusion at bases in mature individuals (Fig 231d). Pterostigmata dark brown.

Similar species
Both sexes
See Dragonfly Group 6.
Downy Emerald Dragonfly (p 189, **192**).
Northern Emerald Dragonfly (p **192**, 194).

Behaviour/field tips
This rather elusive dragonfly is only active when the sun is shining. As soon as it becomes overcast it will immediately retire to the surrounding tree canopy. The males patrol low over the margins of their breeding waters in determined style, rarely back-tracking or hovering. They are most often seen searching for females in calm, sheltered areas that are deeply shaded beneath overhanging trees. The females can also be found ovipositing in such areas. Whilst so engaged, they hover just above the water, mosses or peat and continuously dip their abdomen towards the surface to lay their eggs. However, most of the adults' lives are spent in woodland away from the water where they mature, hunt and mate. They can then be extremely difficult to find.

White-faced Darter
Leucorrhinia dubia (Vander Linden, 1825)

Flight period
Usually from early May to late August. Optimum from early June to late July.

Distribution and status
The White-faced Darter is a generally scarce and highly localised species.
England: Restricted mainly to the Mosses of Fenn's, Whixall and Bettisfield in Shropshire, Chartley in Staffordshire and Scaleby and Foulshaw in Cumbria. It is absent from the Isle of Man and the Isles of Scilly.
Scotland: Found at widely scattered localities in a triangular area approximately between Argyll Main in the south-west, South Aberdeenshire in the east and West Ross in the north-west. It has not been recorded from any of the offshore islands.
Wales: Restricted to the Denbighshire part of the Whixall/Fenn's/Bettisfield Moss complex.
Ireland: Absent.

Fig. 161 White-faced Darter. Whixall Moss, Shropshire

Habitat preferences
Restricted to oligotrophic, acid bog-pools amongst, or adjacent to, light woodland. These situations are most often found in Scotland but have inadvertently been reproduced by peat-cutting in some of the English and Welsh Mosses.

Fig. 162 Whixall Moss, Shropshire

White-faced Darter

Description
Length *c*. 35mm; wingspan *c*. 60mm.
Adult male (Fig 277)
Head From above, eyes very dark brown. Frons white. From the side, eyes very dark brown. From the front, entire 'face' distinctively white (Fig 278).
Thorax From above, black with antehumeral stripes red. Base of wings with large, red spots (Fig 277). From the side, black. Segment 2 with an angulated, red bar above two small, red spots. Segment 3 with a large, red patch. Legs black.
Abdomen From above, black. Segments 2 & 3 each with a large, red patch. Segments 5–7 each have a large, elongate, red spot (Fig 277). From the side, upperside spots still visible. Lower half of segment 2 with an additional red patch. Lower half of surface of segment 3 white.
Wings Clear with pterostigmata short, broad and black. Hindwing with prominent black patches at the base. Forewing with black basal area smaller (Fig 277).

Immature male (Fig 279)
All red markings of adult replaced by yellow. 'Face' yellow rather than white (Fig 283b). Eyes with lower half green (Fig 283b).

Female (Fig 280)
Head As in immature male (Fig 283c).
Thorax As in immature male (Figs 279; 256b & c).
Abdomen From above, black. Segment 2 with a prominent rhomboidal, yellow spot above a bold, yellow, horizontal bar. Base of segment 3 with a large, yellow spot (Fig 282c). Segments 4–7 each with a pair of yellow, vertical stripes that are usually conjoined to form a bold, vertical bar (Fig 280). From the side, black with lower half of segment 3 white or pale yellow.
Wings Clear with pterostigmata short, broad and black (Fig 280). Hindwing with dark, brown basal patches. Forewing with basal patches much smaller (Fig 282c).

Similar species
Adult male
None. The black and red markings are unique to this species.
See Dragonfly Group 12. See p 225.

Immature male and female
See Dragonfly Group 13.
Black Darter male (p 226).

Behaviour/field tips
This dragonfly is difficult to follow on the wing as its flight is rapid and unpredictable but there are frequent periods of hovering. The males have favoured perches, such as bare earth, boardwalks and twigs low to the water. These are visited regularly and habitually. The best way to photograph the White-faced Darter is therefore to 'stake out' such features and wait for the insect's arrival. The females spend much of their time away from the breeding pools and are best sought as they hunt and bask amongst nearby trees or hedge-lined tracks.

Broad-bodied Chaser
Libellula depressa Linnaeus, 1758

Flight period
Usually from early May to late August. Optimum from early June to late July.

Distribution and status
England and Wales: With the exception of mountainous areas, widespread and common south of a line approximately between Cumberland in the west and South Northumberland in the east, with a few scattered localities further north. It is absent from the Isle of Man and the Isles of Scilly.
Scotland: Restricted to a few scattered localities as far north as Kirkcudbright in the west and Midlothian in the east.
Ireland: Absent.

Habitat preferences
Pools, lakes and sometimes very slow-flowing waterways with areas of bare shoreline and shallows are the favoured habitats (Fig 164). This is often the first species to colonise newly dug garden ponds.

Description
Length *c*. 45mm; wingspan *c*. 75mm.
Adult male (Fig 240)
Head Eyes and frons entirely dark brown (Fig 245b).
Thorax From above, dark brown with antehumeral stripes broad, blue and outwardly arched (Fig 245b). From the side, entirely dark brown. Legs black.
Abdomen Broad with bright blue pruinesence. At the base, segment 1 brown (Fig 245b). At the tip, segments 9 & 10 black with clear demarcation from

Fig. 163 Broad-bodied Chaser. New Forest, South Hampshire

Fig. 164 Thompson Water, West Norfolk

Broad-bodied Chaser

segment 8. Segments 3–6 each with a pair of large, pale-yellow, lateral spots (Fig 246b). From the side, dorsally flattened, blue above and dark brown below. Lateral, yellow spots from upperside clearly visible.
Wings Clear with extensive dark-brown patches at their bases (Fig 245b). Pterostigmata long and black (Fig 247b).

Immature male (Fig 251)
Head, thorax and wings Similar to adult but thorax generally paler with antehumeral stripes white rather than blue (Fig 255b).
Abdomen From above, bright orange with a fine, dark-brown, central line that broadens towards the tip to give a brown 'tail'. Segment 1 brown. Segments 3–6 each with a pair of large, pale-yellow lateral spots (Fig 251).

Adult female (Fig 250)
Head, thorax and wings Similar to immature male.
Abdomen From above, similar to immature male but much broader. Darkbrown suffusion at tip far less extensive. Anal appendages inwardly pointing or straight rather than outwardly curved (Fig 250).

Over-mature female (Fig 237)
Similar to adult female but abdomen dark brown, often with traces of the blue pruinesence of the adult male.

Similar species
Adult male
See *Dragonfly Group 8*.
Scarce Chaser adult male (p 197, **198**).
Black-tailed Skimmer male (p **198**, 201).
Keeled Skimmer male (p **198**, 202).

Immature male and adult female
See *Dragonfly Group 9*.
Scarce Chaser female and immature male (p **204**).
Black-tailed Skimmer female and immature male (p **204**, 206).
Four-spotted Chaser female (p **204**, 207).

Over-mature female
See *Dragonfly Group 7*.
Four-spotted Chaser male (p **196**).

Behaviour/field tips
This robust and impressive dragonfly is also one of the more obliging to observe and photograph as both sexes settle for long periods on tall vegetation with a sunny aspect. The males adopt favoured vantage points that give an open view of their territories. From here they watch avidly for male interlopers and prospective mates. Flight sorties and clashes are usually short with the insect soon returning to its original perch. Egglaying takes place around the margins of the water, during which a female hovers and dips the tip of her abdomen into the water onto sub-surface plant material. She is often accompanied closely by her mate. Whilst so engaged, wonderful opportunities for 'flight shots' of both sexes will present themselves to the patient photographer. Immatures of both sexes can often be seen hunting along hedgerows and amongst woodland rides and clearings.

Scarce Chaser
Libellula fulva Müller, 1764

Flight period
Usually between mid-May and late July. Optimum from early June to mid-July.

Distribution and status
England: This generally uncommon species has a very localised, but widespread distribution south of a line approximately between Shropshire, Worcestershire and West Gloucestershire in the west and East Norfolk in the east. There is a central belt, approximately between Oxfordshire in the north and North Hampshire and South Wiltshire in the south, from which the Scarce Chaser appears almost to be absent. It is also yet to be found in either North or South Cornwall, the Isles of Scilly or the Isle of Man.
Scotland, Wales and Ireland: Absent.

Habitat preferences
Mature, slow-flowing lowland rivers with an abundance of bankside vegetation and adjacent scrub are the usual situations that host this species. From here it will occasionally colonise nearby stillwaters such as canals, ditches and pools.

Description
Length *c*. 45mm; wingspan *c*. 75mm.
Adult male (Fig 239)
Head Eyes blue and frons black (Fig 245a).
Thorax Slightly downy and almost entirely black (Fig 245a). Legs very dark brown or black.
Abdomen Slender with bright blue pruinesence. At the base, segment 1 dark brown or black (Fig 245a). At the tip,

Fig. 165 Scarce Chaser immature male. Strumpshaw Fen, East Norfolk

segments 7–10 black with precise definition between these and the blue of segment 6 (Fig 246a). Dark patches or lines may be present on the blue segments where the pruinesence has been removed by a female's legs during mating (Fig 166). From the side, blue above and black below.
Wings Mainly clear with pterostigmata short and black (Fig 247a). Forewing with short, black, basal line, sometimes with a short, cream or yellow bar above. Hindwing with dark, basal triangle interrupted by a yellow or orange, horizontal bar (Fig 245a). Tips occasionally with a small, smoky-brown patch.

Immature male (Fig 249)
Head Eyes and frons brown.
Thorax Slightly downy. From above, orange turning black with maturity (Fig 255a). From the side, entirely brown. Legs black.

Scarce Chaser

Fig. 166 Scarce Chaser male abdomen after mating. Pulborough, West Sussex

Abdomen From above, bright orange with a central row of black, triangular or bell-shaped markings that increase in size towards the tip (Fig 249). Each segment with a fine, black lateral line (Figs 249; 255a). From the side, orange above and dark brown below.
Wings Bases with black, triangular patches, those on the hindwings being the larger (Fig 248). Leading edges broadly suffused yellow (Figs 249; 255a; 256a). Tips sometimes with a small, smoky-brown patch. Pterostigmata short and black (Fig 256a).

Female (Fig 248)
Head Eyes dark brown. Frons pale brown.
Thorax and abdomen As in immature male.
Wings As in immature male, but tips always with a large and conspicuous smoky-brown patch (Figs 248; 256a).

Over-mature female (Fig 285)
Markings as in adult female but general colour very dark brown.

Similar species
Adult male
See *Dragonfly Group 8*.
Broad-bodied Chaser adult male (p **197**, 198).
Black-tailed Skimmer adult and sub-adult male (p **197**, 201).
Keeled Skimmer adult and sub-adult male (p **197**, 202).

Female and immature male
See *Dragonfly Group 9*.
Broad-bodied Chaser female and immature male (p **204**).
Black-tailed Skimmer female and immature male (p **204**, 206).

Over-mature female
See *Dragonfly Group 14*.
Black-tailed Skimmer over-mature female (p **228**, 229).
Keeled Skimmer over-mature female (p **228**, 229).
Common Darter over-mature male and female (p **228**, 229).

Behaviour/field tips
On emerging, the adults seek out sheltered areas such as scrub or woodland clearings. Here both females and immature males can be found sunbathing amongst tall vegetation. If disturbed they usually fly only a short distance and can easily be followed. Beds of nettles in such places provide an excellent source of prey, such as aphids and other small inverte-

Scarce Chaser

brates. Older individuals can also sometimes be seen utilising these areas. Mature males spend much of their time guarding their territories from open perches amongst bankside vegetation. From here they will dart out swiftly over the water to investigate intruders. Clashes are usually short but emphatic and, after periods of observational hovering, the dragonfly will usually return to its favoured 'watchtower'. Mating pairs can often be disturbed from tall vegetation along the riverbanks. The female only returns to the water to oviposit. When doing so she hovers, usually close to the bank, and dips her abdomen towards the water's surface, thus releasing her eggs. Her mate will always be hovering in close attendance to ward off competing males.

Four-spotted Chaser
Libellula quadrimaculata **Linnaeus, 1758**

Flight period
Usually from early May to mid-August. Optimum from late May to late July.

Distribution and status
This is a widespread and common species throughout most of the British Isles but has not been reported from the Isles of Scilly.

Habitat preferences
Almost any still or slow-moving open water may provide a home for this species. It is even able to colonise brackish pools near the sea.

Description
Length *c.* 50mm; wingspan *c.* 80mm.
Male (Fig 236)
Head Eyes dark brown. Frons cream or very pale brown.
Thorax Downy and entirely dark brown. Legs dark brown or black.
Abdomen From above, distinctively short and pointed. Mainly brown, darkening

Fig. 167 Four-spotted Chaser. Thompson Water, West Norfolk

towards the tip. Segments 3–7 each with a pair of prominent cream or pale-yellow lateral spots (Fig 236). From the side, upper half with lateral spots visible. Lower half dark brown.
Wings Hindwing with large, dark-brown triangular patch edged above with a

Four-spotted Chaser

yellow or orange bar (Fig 236). Leading edge of forewing tinged yellow. All wings with characteristic black spot at the node. Pterostigmata short and black (Figs 238a & b).
Form *praenubila* Newman (1833) with pterostigma extended into a large, dark, nebulous patch (Fig 238b). Found regularly throughout the species' range.

Female (Fig 245)
Head As in the male.
Thorax Entirely orange-brown. Legs dark brown or black (Fig 255d).
Abdomen Mainly orange-brown, darkening towards the tip. Markings similar to male (Fig 254).
Wings Similar to male but yellow suffusion along leading edges more intense (Fig 255d).

Similar species
Male
See Dragonfly Group 7.
Broad-bodied Chaser female (p **196**).

Female
See Dragonfly Group 9.
Broad-bodied Chaser female and immature male (p **204**, **207**).

Behaviour/field tips
This may be a very common species, but the Four-spotted Chaser is nonetheless a fascinating creature to watch. The males are highly territorial and, from a favoured perch, watch over their domain for intruding males, passing females and suitable prey. The interception clashes with other males are short and sharp; the original territory owner usually being victorious and returning to his same watchpoint. Other forays over the water appear erratic and involve much hovering. As the species is quite tolerant of human presence, wonderful close views and photo-opportunities often arise. When numbers are particularly high, the male's territorial behaviour appears to lose its structure and a bewildering display of aerobatics can then be seen as the insects, in their dozens, jostle, chase and fight in continuous confrontation. These situations are often profitable times for a passing Emperor Dragonfly to whom the Four-spotted Chaser frequently falls prey. Females are far more secretive and, although they can sometimes be seen copulating over the water, they spend much of their time in surrounding scrubby or wooded areas away from the breeding sites. When a female does return to the water, her time there is often brief as she will be pestered continuously by amorous males – despite being guarded closely by her original suitor. On a warm summer's morning, I can think of fewer pleasant ways of passing the time than watching this intriguing dragonfly.

Black-tailed Skimmer
Orthetrum cancellatum (Linnaeus, 1758)

Flight period
Usually from mid-May to early September. Optimum throughout June and July.

Distribution and status
England and Wales: Widespread and often common in lowland areas throughout, though it appears to be absent from the Isle of Man and the Isles of Scilly.
Scotland: Recorded only from the counties of Roxburghshire and Fifeshire.
Ireland: Found mainly in the central belt bordered by West Donegal and Louth in the north, and Limerick and Waterford in the south. It is well established in the Burren limestone district of Clare and South-east Galway. There have been only a few records from elsewhere.

Habitat preferences
The Black-tailed Skimmer can be found at a wide variety of still and slow-flowing waters, including coastal lagoons. Those with a gently sloping gradient are preferred as the water near the edges is shallow and warm and therefore advantageous to larval development. Such sites also provide what appears to be a prerequisite for the species in the form of large areas of bare shoreline or miniature 'beaches' where the adults can bask on the warm substrate. Man-made tracks and boardwalks near breeding sites are also used frequently for this purpose (Fig 169). Similar conditions may be created by the trampling of stock or human activities. Nearby scrub, bushes or hedges are necessary as hunting grounds and sanctuary for newly emerged adults.

Description
Length *c*. 45mm; wingspan *c*. 75mm.
Adult male (Fig 241)
Head Frons brown. Eyes dark green (Fig 245c).
Thorax Slightly downy and almost entirely dark brown. Legs black (Fig 245c).
Abdomen From above, mainly blue with inconspicuous pale-yellow lateral edges in younger individuals. There are often black smudges and scratches on the blue segments where the pruinesence has been removed by the female's legs during mating. At the base, segment 1 dark brown (Fig 245c). At the tip, segments 8–10 black with indistinct border with the blue of segment 7 (Fig 246c). From the side, blue above and brown below (Fig 168).
Wings Clear with yellow-tinged costa and dark-grey or black pterostigmata (Fig 247c).

Immature male (Fig 253) **and female** (Fig 252)
Head From above, frons and eyes pale brown (Fig 255c). From the side, eyes pale brown above and pale green below.
Thorax From above, yellow, edged with dark brown (Fig 255c). From the side, yellow with black intersegmental lines (Fig 170). Legs dark brown.
Abdomen From above, yellowish central stripe edged with a dark brown bar. Each segment with an elongated, yellow, lateral spot (Figs 252; 253). From the side, yellow with dark-brown intersegmental divisions. Dark-brown bars of upperside visible (Fig 170).

Black-tailed Skimmer

Fig. 168 Black-tailed Skimmer. Thompson Water, West Norfolk

Fig. 169 A shingle track at Cley. East Norfolk

Black-tailed Skimmer

Wings Clear with costa usually dark yellow. Pterostigmata black (Fig 256c).

Sub-adult male
Similar to immature male but abdomen with increasing degrees of blue as maturity approaches (Figs 242; 246e).

Over-mature female (Fig 286)
Markings as in adult female but ground colour dark brown, often with traces of the blue pruinesence found in the male.

Similar species
Adult and sub-adult male
See Dragonfly Group 8.
Scarce Chaser adult male (p 197, **201**).
Broad-bodied Chaser adult male (p 198, **201**).
Keeled Skimmer adult and sub-adult male (p **201**, 202).

Female and immature male
See Dragonfly Group 9.
Scarce Chaser female and immature male (p 204, **206**).
Broad-bodied Chaser female and immature male (p 204, **206**).

Over-mature female
See Dragonfly Group 14.
Scarce Chaser over-mature female (p **228**).
Keeled Skimmer over-mature female (p **228**, 229).
Common Darter over-mature female (p **228**, 229).

Fig. 170 Black-tailed Skimmer female. Fakenham, West Norfolk

Behaviour/field tips
At a known locality one should search along the 'beaches' of lakes and pools or man-made tracks as it is here that the male Black-tailed Skimmer spends much of its time basking and protecting fiercely its territory against encroaching males. From a favoured perch, they also await the infrequent visits of females, darting out quickly to check for a prospective partner before returning, if the search was fruitless, to the same watchpoint. Females spend most of their lives hunting amongst nearby scrub and bushes or along hedgerows. They will take prey as large as satyrid butterflies or damselflies. Their visits to the breeding sites are solely for the purpose of mating and laying eggs. The latter is done over shallow water whilst hovering alongside her ever-attentive mate.

Keeled Skimmer
Orthetrum coerulescens (Fabricius, 1798)

Flight period
Usually late May to early September. Optimum from mid-June to early August.

Distribution and status
England: Widespread and locally common south-west of a line approximately between North Somerset in the north and South Hampshire and the Isle of Wight in the south. There are further strongholds centred on the counties of Surrey and Berkshire in the south, West and East Norfolk in the east and Westmorland and Cumberland in the north. There are also outlying colonies in North-east Yorkshire. Elsewhere there are scattered records of this nomadic species throughout the country. It is absent from the Isle of Man and the Isles of Scilly.
Scotland: Very much a westerly distribution between Kirkcudbrightshire in the south and West Ross in the north. It is found on many of the islands of the Inner Hebrides, but it has not been recorded from the Outer Hebrides or the Northern Isles.
Wales: Widespread throughout the western half of the country, including the island of Anglesey.
Ireland: Has been recorded from many scattered localities throughout, but it is widespread and common in suitable localities west of a line between West Cork in the south and East Mayo in the north. There are also eastern strongholds centred on the counties of Wicklow and Down.

Habitat preferences
This species is a denizen of lowland heathland and moorland where it breeds in bog pools, flushes, runnels and small streams.

Description
Length *c*. 40mm; wingspan *c*. 65mm.
Adult male (Fig 243)
Head From above, frons dark brown. Eyes dark bluish-green (Fig 245d). From the side, frons mainly pale brown. Eyes dark bluish-green above, becoming paler below.
Thorax From above, dark brown with antehumeral stripes narrow, straight and cream or pale yellow (Fig 245d). From the side, mainly dark brown. Legs black.

Fig. 171 Keeled Skimmer male in typical 'dropwing' pose.
New Forest, South Hampshire

Keeled Skimmer

Abdomen From above, almost entirely blue with a fine, black, central line (Fig 246d). Some of the blue pruinesence may be removed by the female's legs during mating. From the side, blue above and brown below.
Wings Clear with pterostigmata long and amber in colour (Fig 247d).

Sub-adult male (Fig 244)
Head Frons yellow. Eyes pale brown (Fig 244).
Thorax As in adult male but antehumeral stripes paler (Fig 244).
Abdomen From above, each segment with a pair of long, yellow, vertical panels divided by, and edged with, increasing amounts of blue as maturity approaches (Figs 244; 246f).
Wings As in adult male (Fig 247d).

Adult female (Fig 264)
Head From above, frons pale yellow. Eyes pale brown, turning to greenish-blue with age (Fig 270a). From the side, frons pale yellow. Eyes pale brown, turning to greenish-blue above and greyish-brown below (Fig 273a).
Thorax From above, brown with antehumeral stripes broad and pale cream or white. Top of thorax with two large prominent cream or white, central spots

Fig. 172 Typical Keeled Skimmer habitat. Holt Lowes, West Norfolk

Keeled Skimmer

(Fig 270a). From the side, segment 1 chestnut-brown. Segments 2 & 3 pale brown. Legs pale brown (Fig 273a).
Abdomen From above, pale straw-coloured. Each segment with a black, inverted 'T'-shaped, central marking (Fig 264). From the side, straw-coloured with a series of black, horizontal lines along the lower edge.
Wings Fore-margins tinged yellow. Costa white (Fig 270a). Pterostigmata amber (Fig 276a).

Immature male (Fig 265)
Similar to female but most segments with black lateral bars. Wings lack the yellow-tinged fore-margins.

Over-mature female (Fig 287)
Markings as in adult female but ground-colour very dark grey-brown. The distinctive amber-coloured pterostigmata remain.

Similar species
Adult and sub-adult male
See Dragonfly Group 8.
Scarce Chaser male (p 197, **202**).
Broad-bodied Chaser male (p 198, **202**).
Black-tailed Skimmer adult and sub-adult male (p 201, **202**).

Adult female and immature male
See Dragonfly Group 11.
Common Darter female, immature male and immature form *nigrescens* (p **212**, 214).
Red-veined Darter female and immature male (p **212**, 217).
Black Darter female and immature male (p **212**, 220).
Ruddy Darter female and immature male (p **212**, 222).

Over-mature female
See Dragonfly Group 14.
Scarce Chaser over-mature female (p 228, **229**).
Black-tailed Skimmer over-mature female (p 228, **229**).
Common Darter over-mature female (p **229**).

Behaviour/field tips
Where it is known to occur, this species can be very easy to find as it spends most of its time sunbathing on the ground and 'hopping' between clumps of low vegetation. A walk through heathers or Bog Myrtle (*Myrica gale*) near water is all that is needed to ascertain its presence as they are easily disturbed upon one's approach. Flights are usually short, and the insect can be followed and studied closely so long as care is taken and patience applied. However, obtaining good photographs can be challenging as, when the dragonfly settles, it slowly but surely pushes its wings downward and forward, resulting in an awkward pose (Fig. 171). Therefore, the quicker one can take a shot (i.e. before the wings begin to drop), the more satisfactory the photographs will be.
At the breeding site, both sexes can be seen on their short flight sorties close to the water's margins.

Black Darter
Sympetrum danae (Sulzer)

Flight Period
Usually from late June to mid-October. Optimum during August and September.

Distribution and status
England. Widespread and common in suitable localities west of a line approximately between South-east Yorkshire in the north and South Devon in the south. There are further strongholds centred on the lowland heaths of South Hampshire, Dorset, Surrey, Sussex, Berkshire and West Norfolk. It is present in a few localities in the Isle of Man but absent from the Isles of Scilly.
Scotland: Widespread and often common throughout, including all larger offshore islands except for the Shetland Islands.
Wales: Widespread and often common throughout, including the island of Anglesey.
Ireland: Widespread at scattered localities throughout except for the extreme south-eastern counties where it is scarce.

Fig. 173 Black Darter. Roydon Common, West Norfolk

Black Darter

Habitat preferences

The Black Darter breeds only in shade-free, acidic bog-pools and cut-over peat workings on moorlands and heathland (Fig 174). The abundant growth of marginal emergent sedges and rushes, *Sphagnum* and other bog-mosses are essential elements.

Description

Length *c*. 35mm; wingspan *c*. 60mm.
Adult male (Fig 281)
Head From above, eyes dark brown. Frons black (Fig 282a). From the side, eyes brown above and pale green below (Fig 283a). From the front, frons yellow, edged with black along the top and sides (Fig 283a).
Thorax From above, downy and black (Fig 282a). From the side there are two broad, yellow, diagonal panels divided by a black band. Within this band is a series of three isolated yellow spots. There are further isolated single, yellow spots at the bases of the fore- and mid-legs. Legs black (Fig 283a).
Abdomen From above, distinctly club shaped. Mainly black with a varying number of vertical pairs of yellow or orange, central lines (Fig 281). These

Fig. 174 An East Anglian outpost for Black Darter. Dersingham Bog, West Norfolk

Black Darter

are sometimes absent on segments 4–6 of fully mature adults. From the side, mainly black. Segments 2 & 3 with large, yellow panels. Upper half of segments 7 & 8 each with a single, yellow spot (Fig 173).
Wings Clear with pterostigmata black (Fig 281).

Adult female (Fig 271)
Head From above, eyes brown. Frons yellow (Fig 270d). From the side and front, similar to male (Fig 273d).
Thorax From above, yellow-orange. Segment 1 with a characteristic large, dark-brown, downward-pointing triangle (Fig 270d). From the side, similar to male.
Abdomen From above, straw-coloured. Segments 8 & 9 each with a conspicuous black, approximately triangular spot (Fig 271). From the side, straw-coloured above and mainly black below (Fig 173).
Wings Clear with pterostigmata white, turning to black with maturity (Fig 276f).

Immature male (Fig 272)
Very similar to female but abdomen from above club shaped and segments 5–9 each edged laterally with black.

Similar species
Adult male
See Dragonfly Group 13.
White-faced Darter female and immature male (p **226**).

Female and immature male
See Dragonfly Group 11.
Keeled Skimmer female and immature male (p 212, **220**).
Common Darter female, immature male and immature form *nigrescens* (p 214, **220**).
Red-veined Darter female and immature male (p 217, **220**).
Ruddy Darter female and immature male (p **220**, 222).

Behaviour/field tips
This busy little dragonfly is difficult to follow in flight as its aerial sorties are quick, darting and erratic. However, it perches frequently on bare soil, dead branches and waterside vegetation, and can then often be approached and photographed with confidence. Unmated females can frequently be seen flying quickly over the water's surface where they occasionally attract the attentions of several suitors at the same time. The sight of the ensuing melee is fascinating to watch. Pairs in tandem can easily be followed as they leave the pond to mate in the surrounding low vegetation, thus presenting 'intimate' photo-opportunities.

Red-veined Darter
Sympetrum fonscolombii (Sélys, 1840)

Flight period
Immigrant individuals begin to arrive in the British Isles during late April and usually attempt to breed. At warm southern localities their offspring start to emerge during August and continue to fly until early November. In more northerly climes the larvae resulting from spring matings overwinter and emerge as adults the following spring. The optimum time to look for the species is from late May to mid-August when the numbers of immigrants and those of 'home-bred' individuals are at their highest.

Distribution and status
The Red-veined Darter was once regarded as being a scarce visitor to our shores. However, a significant increase in sightings, and breeding success, recorded every year (sometimes during several subsequent years at the same site), justify classifying it now as a resident species.
England and Wales: Recorded at widely scattered localities throughout, though there is a strong tendency for it to establish colonies at coastal sites and their hinterlands. It has not been reported from Anglesey and there are very few records from the Isle of Man.
Scotland: Found only at a few localities in the south and has never been reported from the offshore islands. There are several recent records from the Isles of Scilly.
Ireland: Sightings appear to have increased in recent times but are usually restricted to coastal localities in the east and south.

Fig. 175 Red-veined Darter. Mallorca

Fig. 176 A breeding site for Red-veined Darter. Kelling Water Meadows, East Norfolk

Habitat preferences
The warm, shallow margins of flooded earth-workings, pools, settling-beds and lakes with sparse emergent vegetation,

Red-veined Darter

and the shallow waters of coastal lagoons and dune slacks provide the required breeding habitats for this species (Fig 77).

Description
Length *c*. 40mm; wingspan *c*. 65mm.
Adult male (Fig 259)
Head From above, eyes dark red. Frons paler (Fig 261b). From the side, eyes dark red above and pale blue below. From the front, frons pink, edged black along the front and sides (Fig 263b).
Thorax From above, dark red and slightly downy (Fig 261b). From the side, the outer two of three pale panels pale grey or pale blue. Central panel pale grey and tinged red. Legs black, edged along the outer surface with pale grey or cream (Fig 263b).
Abdomen From above, red. Segments 8 & 9 each with a small, black, central, vertical bar (Fig 259). From the side, red with a series of dark, horizontal lines along the lower edge (Fig 175).
Wings Clear with veins red along fore-margins. Hindwing with extensive orange suffusion in basal area (Figs 259; 261b 262). Pterostigmata characteristically pale yellow/orange, edged with black along top and bottom (Fig 262).

Female (Fig 268)
Head From above, eyes brown. Frons pale yellow/cream (Fig 270e). From the side, eyes brown above and pale blue below. Frons pale yellow/cream (Fig 273e).
Thorax From above, straw-coloured with antehumeral stripes pale yellow, edged laterally with orange (Fig 270e). Segments 8 & 9 each with a prominent black, central, vertical bar (Fig154). From the side, similar to male but with central segment tinged yellow (Fig 175). Legs as in male.
Abdomen From above, mainly straw-coloured. Segment 1 black. Segments 2 & 3 each with a black, central, vertical line (Fig 270e). Segments 8 & 9 each with a conspicuous black, central, vertical bar (Fig 268). From the side, lower half of each segment with a long, pale-cream rectangular panel edged strongly above and below with black (Fig 175).
Wings Clear with veins yellow along fore-margins. Hindwing with extensive orange suffusion in basal area (Fig 270e). Pterostigmata characteristically pale yellow/orange, edged with black along top and bottom (Fig 268).

Immature male (Fig 269)
Head and wings Similar to female.
Thorax Similar to female but usually without antehumeral stripes.
Abdomen Similar to female but without black markings on segments 1–3.

Similar species
Adult male
See *Dragonfly Group 10*.
Common Darter adult male and form *nigrescens* (p 208, **209**).
Ruddy Darter male (p **209**, 210).

Female and immature male
See *Dragonfly Group 11*.
Keeled Skimmer female and immature

Red-veined Darter

male (p 212, **217**).
Common Darter female, immature male and form *nigrescens* female and immature male (p 214, **217**).
Black Darter female and immature male (p **217**, 220).
Ruddy Darter female and immature male (p **217**, 222).

Behaviour/field tips
As a powerful migrant, the Red-veined Darter may occur just about anywhere. It is often in the company of other similar species with which it also shares very similar patterns of behaviour. Daunting as it may seem then, one must patiently and methodically check the features of every red or straw-coloured darter seen until a positive identification is made.

Common Darter
Sympetrum striolatum (Charpentier, 1840)

Flight period
Usually from late June to late October. Optimum throughout August and September.

Distribution and status
England, Wales and Ireland: Widespread and common throughout except for the highest of ground. It is present in the Isle

Fig. 177 Common Darter. Strumpshaw Fen, East Norfolk

Common Darter

of Man, Anglesey and the Isles of Scilly. *Scotland*: With the exception of the mountainous districts, it is widespread and common in the western half of the country, including the offshore islands. It is scarce or absent from the eastern counties of North and South Aberdeenshire, Kincardineshire and their surrounding areas. It has not yet been recorded from the northern Isles of Orkney and Shetland Islands.

Habitat preferences
This species will colonise most habitat types, including the brackish conditions of coastal localities.

Description
Length *c*. 40mm; wingspan *c*. 60mm.
Adult male (Fig 257)
Head From above, eyes dark reddish-brown (Fig 261a). From the side, eyes dark reddish-brown above and pale greenish-brown below. From the front, frons pale brown, edged black along the top (Fig 263a).
Thorax From above, brown, slightly downy and with antehumeral stripes pale yellow or cream (Fig 261a). From the side there are three large, diagonal panels. Outer two yellow; central panel brown, sometimes with a line of three pale, central spots (Fig 263a). Legs dark brown with a characteristic pale yellowish-brown stripe along the outer surface of each (Fig 263a).
Abdomen From above, mainly brick-red. Segments 8 & 9 each with a small, black, central, triangular marking (Fig 257). Younger adults often with a pair of pale-orange, square or rectangular lateral patches on each segment. From the side, ventral surface with a series of dark-brown or black, horizontal bars (Fig 177).
Wings Clear with pterostigmata dark brown (Fig 257).

Adult female and immature male (Fig 266)
Head From above, eyes brown (Fig 270b). From the side, eyes brown above and greenish-yellow below. From the front, frons pale yellow, edged black along the top (Fig 273b).
Thorax From above, pale brown with antehumeral stripes absent or inconspicuous and straw-coloured (Fig 270b). From the side, including legs, similar to male (Fig 273b).
Abdomen From above, mainly straw-coloured with fine, brown intersegmental divisions. Segments 8 & 9 each with a small, black, central, vertical, triangular bar (Fig 266). From the side, ventral surface with a series of black, horizontal lines (Fig 177).
Wings Clear with pterostigmata dark brown. Female with small basal patch tinged orange.

Over-mature female (Fig 288)
As above but with head, thorax and abdomen dark smoky-brown. Dark markings more extensive and antehumeral stripes more pronounced.

Common Darter

Fig. 178 'Highland' Darter. Isle of Mull

Form *nigrescens* (Lucas) 'Highland Darter' (Figs 258; 267; Fig 178)
This taxon was long thought to be a species in its own right (*Sympetrum nigrescens* Lucas, 1912) that was restricted to north-western Scotland and the adjacent off-shore islands. However, recent DNA studies by Pilgrim & Von Dohlen (2007) and Parkes *et al.* (2009) conclude that it is merely a melanic form of Common Darter that presumably utilises its generally darker colouration for more efficient absorption of heat whilst perching on warm surfaces. This is a trait seen commonly in insects from northern latitudes. The genetic and morphological variability found in the above works was such that not even the criteria for subspecific status could be satisfied.

Description
Similar in all respects to the typical form but all of the dark markings are deeper in shade and more extensive. Both sexes with conspicuous straw-coloured antehumeral stripes.

Similar species
Adult male (typical and form *nigrescens*)
See Dragonfly Group 10.
Red-veined Darter male (p **208**, 209).
Ruddy Darter male (p **208**, 210).

Female and immature male (typical and form *nigrescens*)
See Dragonfly Group 11.
Keeled Skimmer female and immature male (p 212, **214**).
Red-veined Darter female (p **214**, 217).
Black Darter female and immature male (p **214**, 220).
Ruddy Darter female and immature male (p **214**, 222).

Over-mature female
See Dragonfly Group 14.
Scarce Chaser female (p 228, **229**).
Black-tailed Skimmer (p 228, **229**).
Keeled Skimmer (p **229**).

Behaviour/field tips
The observer should have no problems in finding this widespread, common and sometimes abundant species.

Ruddy Darter
Sympetrum sanguineum (Müller, 1764)

Flight period
Usually from early June to mid-October. Optimum throughout July and August.

Distribution and status
England: Widespread and generally common throughout except for West and East Cornwall and North and South Devon. Here it is relatively scarce. Absent from the Isle of Man. Recorded recently from the Isles of Scilly.
Scotland: Absent, or almost so; the only confirmed record being from Berwickshire in 2003 (Cham *et al.*, 2014).
Wales: Present in the counties of the east and those of the extreme south and south-west, with a few scattered records from elsewhere. It appears to be absent from the island of Anglesey.
Ireland: Widespread and generally common throughout a central belt approximately between Fermanagh and Down in the north and Clare and Wexford in the South. North of this area there are very few records, but to the south it is well established in several of the extreme south-western counties.

Habitat preferences
The Ruddy Darter favours older established sites than the Common Darter but, like its relative, it can be found in a wide variety of still or very slow-flowing habitats. Warm, shallow

Fig. 179 Ruddy Darter mating pair. Thompson Common, West Norfolk

Ruddy Darter

waters with dense emergent and marginal vegetation are preferred.

Description
Length *c*. 25mm; wingspan *c*. 55mm.
Adult male (Fig 260)
Head From above, eyes deep red (Fig 261c). From the side, eyes deep red above and dark greenish-brown below. From the front, frons pink, edged black along the top and sides (Fig 263c).
Thorax From above, brown and slightly downy. Antehumeral stripes absent (Fig 261c). From the side, outer two segmental panels pale pink. Legs entirely black (Fig 263c).
Abdomen From above, mainly blood-red and characteristically club shaped. Segments 8 & 9 each with a black, central, vertical bar (Fig 260). From the side, ventral surface with a series of black, horizontal lines.
Wings Clear with pterostigmata dark brown (Fig 260). Basal area with very small, orange or red patch (Fig 261c).

Female (Fig 274)
Head From above, eyes brown (Fig 270c). From the side, eyes brown above and greenish-yellow below. From the front, frons yellow, edged black along the top and sides (Fig 273c).
Thorax From above, brown and slightly downy. Antehumeral stripes absent (Fig 270c). From the side, yellowish-brown with bold, black intersegmental divisions. Legs entirely black (Fig 273c).
Abdomen From above, straw-coloured with fine, black intersegmental divisions. Segments 8 & 9 each with a black, central, vertical bar (Fig 274). From the side, lower half with a double series of black, horizontal lines.
Wings Clear with pterostigmata grey, turning dark brown with maturity. Basal area with small patch suffused yellow (Figs 274; 249c).

Immature male (Fig 275)
Similar to female but abdomen is distinctly club shaped (Fig 275).

Similar species
Adult male
See Dragonfly Group 10.
Common Darter adult male and form *nigrescens* adult male (p 208, **210**).
Red-veined Darter male (p 209, **210**).

Female and immature male
See Dragonfly Group 11.
Keeled Skimmer female and immature male (p 212, **222**).
Common Darter female, immature male and form *nigrescens* female and immature male (p 214, **222**).
Red-veined Darter female and immature male (p 217, **222**).
Black Darter female and immature male (p 220, **222**).

Behaviour/field tips
Within its distributional range, this very common species should be no problem to find. As with all of its compatriots, one should simply look closely at each red or straw-coloured darter seen and, eventually, a Ruddy Darter will be found.

CHAPTER FOUR
Dragonflies
Identification of the resident species

Dragonfly Group 1
Large (wingspan *c.* 90mm)
Abdomen brown   p 172

Fig. 180
Norfolk Hawker

Dragonfly Group 2
Large (wingspan *c.* 90mm)
Abdomen with black
and gold markings   p 174

Fig. 181
Golden-ringed Dragonfly

Dragonfly Group 3
Large (wingspan *c.* 90mm)
Abdomen with blue
or lilac markings   p 174

Fig. 182
Migrant Hawker

Dragonfly Group 4
Large (wingspan *c.* 90mm)
Abdomen with green
or yellow markings   p 183

Fig. 183
Southern Hawker

Dragonfly Group 5
Medium-sized (wingspan *c.* 65mm)
Abdomen 'club' shaped with
yellow or pale-green markings   p 188

Fig. 184
Club-tailed Dragonfly

Dragonfly Group 6
Medium-sized (wingspan *c.* 70mm)
Abdomen metallic bronze-green   p 189

Fig. 185
Brilliant Emerald

Dragonfly Group 7
Medium-sized (wingspan *c.* 70mm)
Abdomen dull brown   p 196

Fig. 186
Four-spotted Chaser

Dragonfly Group 8
Medium-sized (wingspan *c.* 70mm)
Abdomen blue p 197

Fig. 187
Broad-bodied Chaser male

Dragonfly Group 9
Medium-sized (wingspan *c.* 70mm)
Abdomen orange or yellow p 204

Fig. 188
Broad-bodied Chaser female

Dragonfly Group 10
Small (wingspan *c.* 55mm)
Abdomen red p 208

Fig. 189
Ruddy Darter male

Dragonfly Group 11
Small (wingspan *c.* 55mm)
Abdomen straw-coloured p 212

Fig. 190
Common Darter female

Dragonfly Group 12
Small (wingspan *c.* 55mm)
Abdomen black and red p 225

Fig. 191
White-faced Darter male
© Keith Mallett

Dragonfly Group 13
Small (wingspan *c.* 55mm)
Abdomen black and yellow p 226

Fig. 192
Black Darter male

Dragonfly Group 14
Small to medium-sized
(wingspan *c.* 55-70mm)
Abdomen smoky-brown p 228

Fig. 193
Keeled Skimmer over-mature

Dragonfly Group 1
Large, abdomen brown

Two species: Brown Hawker, Norfolk Hawker

Fig. 194 Brown Hawker male
East Norfolk

Fig. 195 Brown Hawker female
East Suffolk

Fig. 196 Norfolk Hawker male
East Norfolk

Fig. 197 Norfolk Hawker female
East Norfolk

Brown Hawker
Aeshna grandis (Linn.)

Male (Fig 194)
Female (Fig 195)
For species account see p 119.

Similar species
The only species with which the Brown Hawker may be confused is the Norfolk Hawker. It differs in the following ways:

1) Wings heavily tinged orange-brown rather than clear (Fig 194).
2) From above, thorax and abdomen of male with blue spots (Fig 194).
3) From the side, thorax of both sexes with a pair of conspicuous rather than dull, pale bars (Fig 199a).
4) From above, abdomen of both sexes with pairs of small, whitish dots (Figs 194; 195).
5) From the side, abdomen of both sexes with a series of conspicuous pale,

Dragonfly Group 1

horizontal spots (Fig 199a).
6) Eyes brown or blue rather than bright green (Figs 198a; 199a).

Norfolk Hawker
Anaciaeschna isosceles (Müller)

Male (Fig 196)
Female (Fig 197)
For species account see p 127.

Similar species
The only other dragonfly this species may be confused with is the Brown Hawker. It differs in the following ways:
1) Wings clear rather than heavily tinged orange-brown (Figs 196; 197).
2) From above, thorax and abdomen without blue spots (Figs 196; 197).
3) From the side, thorax of both sexes with a pair of dull cream rather than conspicuous whitish-yellow bars (Fig 199b).
4) From above, segment 2 of the abdomen with conspicuous pale-yellow, isosceles-shaped triangular marking (Fig 199b).
5) Eyes bright green rather than brown or blue (Fig 198b).
6) From above, abdomen of both sexes without conspicuous pairs of small, whitish dots (Figs 196; 197).
7) From the side, abdomen of both sexes without a series of conspicuous pale, horizontal spots (Fig 199b).

Fig. 198 Dorsal view of thorax and base of abdomen a. Brown Hawker b. Norfolk Hawker

Fig. 199 Lateral view of thorax and base of abdomen a. Brown Hawker b. Norfolk Hawker © Gary Thoburn

Dragonfly Group 2
Large, abdomen with black and gold markings
One species: Golden-ringed Dragonfly

Fig. 200 Golden-ringed Dragonfly male
South Hampshire

Fig. 201 Golden-ringed Dragonfly female
Isle of Mull

Golden-ringed Dragonfly
Cordulegaster boltonii (Donovan)

Male (Fig 200)
Female (Fig 201)

For species account see p 137.

Similar species
None. Both sexes of the Golden-ringed Dragonfly are unlike any other species found in the British Isles.

Dragonfly Group 3
Large, abdomen with blue or lilac markings
Six species: Hairy Dragonfly male, Emperor Dragonfly male, Azure Hawker male and female 'blue' form, Common Hawker male and female 'blue' form, Migrant Hawker adult and immature male, Southern Hawker immature male
Note: See also Southern Migrant Hawker in Chapter Five (p 230).

Hairy Dragonfly
Brachytron pratense (Müller)

Male (Fig 202)
For species account see p 131.

Similar species
Because the Hairy Dragonfly is on the wing so early in the year, it is unlikely to be seen in the company of other blue-spotted hawkers. This, and the distinctly downy appearance of the thorax (Fig

Dragonfly Group 3

211a), should preclude confusion with the other species in this group.

Emperor Dragonfly
Anax imperator Leach

Male (Fig 203)
For species account see p 129.

Similar species
This true 'Emperor' of dragonflies is large and very robust in build. Its impressive physique, the entirely green side of the thorax and bright-blue abdomen with its distinctive black, central line set it apart in appearance from any other British dragonfly.

Azure Hawker
Aeshna caerulea (Ström)

Male (Fig 204)
Female 'blue' form (Fig 205)
For species account see p 114.

Similar species
Confusion with some of the other species in this group is unlikely for the following reasons:
Hairy Dragonfly flies much earlier in the year. This, their different geographical ranges and habitat preferences and the downy appearance of the Hairy Dragonfly's thorax should preclude confusion with the present species.
Emperor Dragonfly and Migrant Hawker are absent from the Scottish Highlands. This leaves the male and the female 'blue'

Fig. 202 Hairy Dragonfly male
West Norfolk

Fig. 203 Emperor Dragonfly male
West Norfolk

Fig. 204 Azure Hawker male
West Ross © Graham Sherwinn

Dragonfly Group 3

form of the Common Hawker and the immature, blue-spotted phase of the male Southern Hawker with which it can be confused. From these it differs in the following ways:

• **Male and female 'blue' form from male and female 'blue' form of Common Hawker**
1) From above, male thorax with antehumeral stripes short and parallel rather than long and divergent (Fig 211c).
2) From the side, thorax with blue bars narrow and angulated rather than broad and straight (Fig 212c).
3) From above, abdomen of both sexes with two pairs of blue spots on each segment rather than a pair of blue spots and a pair of small, yellow triangles (Figs 204; 205).
4) From the side, male abdomen with segment 3 almost entirely blue rather than blue below and brown above (Fig 212c).
5) From above, contact between eyes short rather than long (Figs 211c & d).
6) Wings with costa only faintly rather than conspicuously yellow. Note that this feature varies in intensity in both species.

• **Male and female 'blue' form from immature male Southern Hawker**
1) Absent from England and Wales (see species account p 114).
2) Very unlike to be met with on Azure Hawker breeding grounds, but it is conceivable the two might be encountered hunting in more wooded areas at slightly lower altitudes.
3) From above, thorax with antehumeral stripes absent, or almost so, rather than large and broad (Figs 211c & d).
4) From the side thorax with bars separate, narrow and pale blue rather than broad, pale yellow and merged into a marbled pattern (Figs 212c & d).
5) From above, abdomen with two pairs of blue spots on each segment rather than a pair of small, blue triangles above a pair of blue spots (Figs 204; 205).
6) From above, abdomen lacking conspicuous pale-yellow 'golf tee'-shaped marking on segment 2 (Figs 211c & d).
7) From above, abdomen with separate pairs of blue spots on segments 9 & 10 rather than conjoined or coalesced into a bar (Fig 204).
8) Head with contact between eyes short rather than long (Figs 211c & d).

Fig. 205 Azure Hawker female 'blue' form West Ross © Graham Sherwin

Common Hawker
Aeshna juncea (Linn.)

Male (Fig 206)
Female 'blue' form (Fig 207)

Dragonfly Group 3

Fig. 206 Common Hawker male West Ross

Fig. 207 Common Hawker female 'blue' form. West Ross

For species account see p 121.

Similar species

Because the Hairy Dragonfly is usually on the wing earlier in the year, it is unlikely to be found in company with the present species. Common Hawker also lacks the distinctly downy thorax of the Hairy Dragonfly. The unspotted abdomen of the Emperor Dragonfly precludes confusion with the present species. The Common Hawker is similar in appearance to the remaining species in this group but differs in the following ways:

• **Male and female 'blue' form from male and female 'blue' form Azure Hawker**
1) Azure Hawker absent from England, Wales and Ireland (see species account p 114).
2) From above, male thorax with antehumeral stripes long and divergent rather than short and parallel (Fig 211e).
3) From the side, thorax with blue bars broad and straight rather than slender and angulated (Figs 212e & f).
4) From above, abdominal segments each with a pair of blue spots beneath a pair of yellow triangles rather than two pairs of blue spots (Figs 206; 207).
5) From the side, male abdomen with segment 3 brown above and blue below rather than mainly blue (Fig 212e).
6) From above, head with contact between eyes long rather than short (Figs 211e & f).
7) Wings with costa strongly rather than faintly yellow, though this varies in intensity (Fig 212e).

• **Male and female 'blue' form from adult male Migrant Hawker**
1) Found throughout Scotland and Ireland. Migrant Hawker absent from most of the former and of very restricted distribution in the latter (see species account p 124).
2) From above, base of male abdomen 'waisted' in appearance rather than straight (Fig 211e).

Dragonfly Group 3

3) From above, thorax with antehumeral stripes long, slender and divergent rather than short and broad (Fig 211e).
4) From the side, thorax with slender, pale-blue stripes rather than broad and yellowish-green (Figs 212e & f).
5) From above, abdomen lacking yellow 'golf tee'-shaped marking above a conspicuous blue band on segment 2 (Fig 211e).
6) From the side, abdomen lacking conspicuous yellow, elongated spot on segment 1 (Fig 212e).
7) Wings with costa usually yellow (Fig 212e).

- **Male and female 'blue' form from immature male Migrant Hawker**
Features above still relevant but:
1) Colours more lilac than blue in immature Migrant Hawker (see Fig 209).
2) Side of thorax with slender, pale-blue stripes rather than broad and pale blue with diffuse edges (Figs 212e & f).
3) Feature 7 above is not yet a useful feature for separation.

- **Male and female 'blue' form from immature male Southern Hawker**
1) Found throughout Scotland and Ireland. Southern Hawker is of restricted distribution in the former and absent from the latter (see species account p 116).
2) From above, thorax with antehumeral stripes absent or indistinct, divergent and pale blue rather than large, broad and yellow (Figs 211e & f).
3) From the side, thorax with two distinct slender, pale-blue stripes rather than bold, yellow-marbled markings (Figs 212e & f).

4) From above, abdomen with blue spots small rather than large and bold (Figs 206; 207).
5) From above, abdomen with spots on segments 9 & 10 separate rather than being conjoined or coalesced into a bar (Figs 206; 207).

Migrant Hawker
Aeshna mixta Latreille

Adult male (Fig 208)
Immature male (Fig 209)
For species account see p 124.

Similar species
The downy thorax and earlier flight period of the Hairy Dragonfly and the restriction in distribution of the Azure Dragonfly to the Scottish Highlands (from which the Migrant Hawker is absent) preclude their confusion with the present species. The unspotted blue abdomen of the Emperor Dragonfly is

Fig. 208 Migrant Hawker adult male West Norfolk

Dragonfly Group 3

unlike that of any other hawker. The Migrant Hawker can be separated from the other species in this group in the following ways:

- **Adult and immature male from male and 'blue' female Common Hawker**

1) Absent from most of Scotland and very locally distributed in Ireland rather than being widespread in both (see species account p 124).
2) Base of abdomen not 'waisted' as in male Common Hawker (Fig 211g).
3) From above, thorax with antehumeral stripes conspicuous, short and broad rather than absent or long and slender (Fig 211g).
4) From the side, adult thorax with broad, yellowish-green stripes rather than slender and blue (Fig 212g).
5) From the side, immature thorax with broad, blue stripes with diffuse edges rather than slender and sharply edged (Fig 212h).
6) From above, abdomen with a conspicuous yellow 'golf tee'-shaped marking above a broad, blue band (Fig 211g).
7) From the side, mature male abdomen with conspicuous yellow, elongated spot on segment 1 (Fig 212g).
8) Wings without yellow costa.

- **Adult and immature male from immature male Southern Hawker**

1) From above, thorax with antehumeral stripes short rather than long and very bold (Fig 211g).
2) From the side, thorax in adult with two clearly divided yellowish-green stripes rather than a marbled pattern of bars and spots (Fig 212g).
3) From the side, thorax in immature with pale-blue markings rather than green (Fig 212h).
4) From above, abdomen with solid, blue bar on segment 2 beneath yellow triangle rather than a pair of conjoined blue spots (Fig 211g).
5) From above, abdomen with spots on segments 9 & 10 small and separate rather than large and conjoined (Figs 208; 209).
6) Wings with pterostigmata brown rather than black (Fig 208).

Southern Hawker
Aeshna cyanea (Müller)

Immature male (Fig 210).
For species account see p 116.

Similar species
The early flight period of the Hairy Dragonfly makes it unlikely that the two species would be seen flying together.

Fig. 209 Migrant Hawker immature male West Norfolk

Dragonfly Group 3

Fig. 210 Southern Hawker immature male West Norfolk

However, in that event the smaller size and distinctly downy thorax of that species should prevent confusion. The distinctive blue abdomen of the Emperor Dragonfly, with its black, central line, is diagnostic of that species.

The immature male Southern Hawker is similar in appearance to the other species in this group but differs in the following ways:

• **From male and female 'blue' form of Azure Hawker**

1) Widespread throughout England and Wales rather than being absent from both (see species account p 116).
2) Unlikely to be met with on each other's breeding grounds, but their hunting areas may conceivably overlap in and around wooded areas adjacent to high moorlands.
3) From above, thorax with antehumeral stripes bold and conspicuous rather than faint or absent (Fig 211i).
4) From the side, thorax with bars yellow, broad and partly merged to form a marbled pattern rather than narrow, separate and pale blue (Fig 212i).
5) From above, abdomen with a pair of small, blue triangles above a pair of blue spots on each segment rather than two pairs of blue spots (Fig 210).
6) From above, abdomen with a conspicuous pale-yellow triangular mark on segment 2 (Fig 211i).
7) From above, abdomen with spots on segment 9 conjoined and on segment 10 coalesced into a bar rather than being separate (Fig 210).
8) From above, head with contact between eyes long rather than short (Fig 211i).

• **From male and female 'blue' form of Common Hawker**

1) Of restricted distribution in Scotland and absent from Ireland rather than being widespread throughout both (see species account p 116).
2) From above, thorax with antehumeral stripes broad and yellow rather than absent or fine and blue (Fig 211i).
3) From the side, thorax with bold, yellow markings rather than two distinct blue bars (Fig 212i).
4) From above, abdomen with blue spots large and bold rather than small (Fig 210).
5) From above, abdomen with spots conjoined on segment 9 and coalesced into a bar on segment 10 rather than being separate on both (Fig210).

• **From immature and adult male Migrant Hawker**

1) From above, thorax with antehumeral

Dragonfly Group 3

Fig. 211 Dorsal view of head, thorax and base of abdomen
a. Hairy Dragonfly male, b. Emperor Dragonfly male, c. Azure Hawker male © Graham Sherwin, d. Azure Hawker female 'blue' form © Graham Sherwin, e. Common Hawker male, f. Common Hawker female 'blue' form, g. Migrant Hawker adult male, h. Migrant Hawker immature male, i. Southern Hawker immature male

Dragonfly Group 3

stripes long and bold rather than short (Fig 211i).
2) From the side, thorax with marbled pattern of yellow bars and spots rather than two distinct bars (Fig 212i).
3) From above, abdomen with two conjoined spots on segment 2 beneath yellow triangle rather than a solid blue band (Fig 211i).
4) From above, abdomen with spots on segments 9 & 10 large and conjoined rather than small and separate (Fig 210).
5) Wings with pterostigmata black rather than brown (Fig 210).

Fig. 212 Lateral view of head, thorax and base of abdomen
a. Hairy Dragonfly male, b. Emperor Dragonfly male, c. Azure Hawker male © Graham Sherwin, d. Azure Hawker female 'blue' form © Graham Sherwin, e. Common Hawker male, f. Common Hawker female 'blue' form, g. Migrant Hawker adult male, h. Migrant Hawker immature male, i. Southern Hawker immature male

Dragonfly Group 4
Large, abdomen with green or yellow markings

Six species: Hairy Dragonfly female, Emperor Dragonfly female, Azure Hawker female 'yellow' form, Common Hawker female 'yellow/green' form, Southern Hawker adult male and female, Migrant Hawker female

Hairy Dragonfly
Brachytron pratense (Müller)

Female (Fig 213)
For species account see p 131.

Similar species
Because of its early flight period (usually early May to mid-June) the Hairy Dragonfly is unlikely to be found flying alongside other similarly marked species. This, its habitat preferences, and the distinctly downy thorax should preclude confusion with the other species in this group.

Emperor Dragonfly
Anax imperator Leach

Female (Fig 214)
For species account see p 129.

Similar species
The female Emperor Dragonfly's large size, unmarked apple-green thorax and the lack of spotted markings on the abdomen preclude confusion with other species in this group.

Azure Hawker
Aeshna caerulea (Ström)

Female 'yellow' form (Fig 215)
For species account see p 114.

Similar species
Of the three other yellow-spotted species in this group two can be discounted immediately as both are absent from the Scottish Highlands. The Hairy Dragonfly flies earlier in the year (usually from late April to late June) and the Migrant Hawker flies much later (usually during

Fig. 213 Hairy Dragonfly female
East Norfolk

Fig. 214 Emperor Dragonfly female
West Norfolk

Dragonfly Group 4

August and September). However, the Azure Hawker is often found in the Highland haunts of the Common Hawker but differs from the 'yellow/green' female of that species in the following ways:

1) Extremely localised and restricted to certain areas of the Scottish Highlands (see species account p 114) rather than being widespread throughout Scotland.
2) Absent from England, Wales and Ireland.
3) From above, thorax with pterostigmata often faintly present rather than always absent (Fig 220c).
4) From the side, thorax with slender, angulated stripes rather than broad and straight (Fig 221c).
5) From above, abdomen with yellow spots small and tinged beige rather than large and tinged green (Fig 215).
6) From above, abdomen with yellow markings on segments 9 & 10 square rather than rounded (Fig 215).
7) From above, eyes with short contact rather than long (Fig 220c).

Fig. 215 Azure Hawker female 'yellow' form. Sweden © Fons Peels, DragonflyPix

Fig. 216 Common Hawker female 'yellow/green' form. West Ross

Common Hawker
Aeshna juncea (Linn.)

Female 'yellow/green' form (Fig 216)
For species account see p 121.

Similar species
The female Hairy Dragonfly flies earlier in the year and has brown tinged rather than clear wings. The female Emperor Dragonfly is mainly green and lacks the spotted pattern on the abdomen, and the large, pure green spots on the abdomen of the Southern Hawker should immediately prevent confusion between those species and this form of the Common Hawker.

The 'yellow/green' form of the present species may be confused with the 'yellow' form of the female Azure Hawker and the female Migrant Hawker but it differs in the following ways:

• **From female 'yellow' form of Azure Hawker**

Dragonfly Group 4

Fig. 217 Southern Hawker adult male
West Norfolk

Fig. 218 Southern hawker female
West Norfolk

1) Widespread in Scotland rather than being very localised (for species account see p 121).
2) Present in England, Wales and Ireland.
3) From above, thorax with antehumeral often present (Fig 220d).
4) From the side, thorax with diagonal bands straight rather than angulated (Fig 221d).
5) From above, abdomen with yellow spots tinged green rather than beige (Fig 216).
6) From above, abdomen with yellow spots on segments 9 & 10 rounded rather than square (Fig 216).
7) From above, head with contact between eyes long rather than short (Fig 220d).
• **From female Migrant Hawker**
1) Found throughout Scotland and Ireland. Migrant Hawker absent from the former and of very restricted distribution in the latter (see species account p 124).
2) From above, thorax with antehumeral stripes sometimes absent (Fig 220d).
3) From the side, thorax with stripes slender and linear rather than broad and rounded (Fig 221d).
4) From above, abdomen with yellow spots tinged green rather than yellow (Fig 216).
5) From above, abdomen with pale markings not broadly edged with black (Fig 216).
6) From above, abdomen lacking conspicuous yellow, 'golf tee'-shaped marking on segment 2 (Fig 220d).
7) From the side, side of abdomen lacking bold, yellow patch on segment 2 (Fig 221d).

Fig. 219 Migrant Hawker female
West Norfolk

Dragonfly Group 4

8) From above, claspers relatively short (Fig 216).
9) Wings with at least base of costa usually tinged yellow (Fig 220d).

Southern Hawker
Aeshna cyanea (Müller)

Adult male (Fig 217)
Female (Fig 218)
For species account see p 116.

Similar species
None. The blue markings towards the tip of the male abdomen are unique to this species. The female is the only species in this group with large, pure green spots on the abdomen.

Migrant Hawker
Aeshna mixta Latreille

Female (Fig 219)
For species account see p 124.

Similar species
Four of the other five species in this group can immediately be discounted as

Fig. 220 Dorsal view of head, thorax and base of abdomen
a. Hairy Dragonfly female, b. Emperor Dragonfly female, c. Azure Hawker female 'yellow' form © Fons Peels, DragonflyPix, d. Common Hawker female 'yellow/green' form, e. Southern Hawker adult male, f. Migrant Hawker female

Dragonfly Group 4

candidates for confusion with the Migrant Hawker. Hairy Dragonfly flies much earlier in the year (see species account p 131). The Southern Hawker and the female Emperor Dragonfly have abdominal markings that are green rather than yellow, and the Azure Hawker is restricted to the Scottish Highlands from where Migrant Hawker is absent (see species account p 114).

This leaves the 'yellow/green' form of the female Common Hawker from which it differs in the following ways:

1) From above, thorax with antehumeral stripes always present (Fig 220f).
2) From the side, thorax with two broad, yellow panels rather than slender stripes (Fig 221f).
3) From above, abdomen with very distinctive 'golf-tee'-shaped marking on segment 2 (Fig 220f).
4) From above, abdomen with yellow spots edged broadly with black and not tinged green (Fig 219).
5) From the side, abdomen with broad, yellow panel on segment 2 (Fig 221f).
6) Wings with costa not tinged yellow (Fig 220f).

Fig. 221 Lateral view of head, thorax and base of abdomen
a. Hairy Dragonfly female, b. Emperor Dragonfly female, c. Azure Hawker female 'yellow' form © Fons Peels, DragonflyPix, d. Common Hawker female 'yellow/green' form, e. Southern Hawker female, f. Migrant Hawker female

Dragonfly Group 5
Medium-sized, abdomen 'club' shaped with yellow or pale-green markings

One species: Club-tailed Dragonfly

Club-tailed Dragonfly
Gomphus vulgatissimus (Linn.)

Male (Fig 222)
Female (Fig 223)
For species account see p 134.

Similar species
None. The separation of the eyes (similar to those of a damselfly) is a unique feature amongst the British dragonflies. Immature males can be separated from females by the sharply indented anal angle of the hindwing (Fig 224b) and, in lateral view, the prominent secondary genitalia at the base of the abdomen (Figs 224a & c).

Fig. 222 Club-tailed Dragonfly male
Oxfordshire

Fig. 223 Club-tailed Dragonfly female
Oxfordshire © Gary Thoburn

Fig. 224 Club-tailed Dragonfly male a. Lateral view, b. Anal corner of hindwing, c. Head and secondary genitalia

Dragonfly Group 6
Medium-sized, abdomen metallic bronze-green

Three species: Downy Emerald Dragonfly, Brilliant Emerald Dragonfly, Northern Emerald Dragonfly

Note: See also Southern Migrant Hawker in Chapter Five (p 230).

Downy Emerald Dragonfly
Cordulia aenea (Linn.)

Male (Fig 225)
Female (Fig 226)
For species account see p 138.

Similar species
The Downy Emerald Dragonfly resembles closely the other two species in this group but differs in the following ways:
- **From both sexes of Northern Emerald Dragonfly**
1) Thorax intensely green rather than slightly tinged blue (Fig 231a).

Fig. 225 Downy Emerald Dragonfly male Berkshire

Fig. 226 Downy Emerald Dragonfly female West Norfolk

Fig. 227 Brilliant Emerald Dragonfly male East Inverness-shire

Fig. 228 Brilliant Emerald Dragonfly female. Surrey © Steve Cham

Dragonfly Group 6

2) Thorax robust rather than slender (Fig 231a).
3) From above, thorax lacking yellow triangle emanating from bases of forewings and pale spot near head (Fig 231a).
4) Thorax and abdominal segments 1 & 2 more conspicuously downy (Fig 231a).
5) Wings with pterostigmata black rather than brown.
6) Wings with bases suffused heavily with orange (Fig 231a).
7) From the front, head lacking yellow spot on each side of the frons (Fig 235a).

• **From male Northern Emerald Dragonfly**
1) From above, abdomen more club shaped with 'waist' shorter (Fig 225).
2) From above, superior claspers short and outwardly curved rather than long and pincer shaped (Fig 233a).
3) Flight patrols usually low to the water's surface.

• **From female Northern Emerald Dragonfly**
1) From above, abdomen lacking yellow spots on segment 3 (Fig 231a).
2) From the side, abdomen lacking yellow spot on lower half of segment 3 (Fig 232a).
3) From the side, abdomen with large, yellow or orange patch on segment 2 rather than cream or yellow lines forming the outline shape of a Wellington boot (Fig 232a).
4) From above, abdomen with claspers stout rather than slender (Fig 233d).
5) From the side, abdomen with vulvar scale beneath segment 8 not usually visible. In Northern Emerald this protrudes at a slight angle from the body (Fig 234a).

• **From both sexes of Brilliant Emerald Dragonfly**
1) Thorax bronze-green rather than bright emerald-green (Fig 232a).
2) From above, thorax lacking fine, yellow or cream, upwardly pointing triangular marking emanating from forewing bases and pale spot near head (Fig 231a).

Fig. 229 Northern Emerald Dragonfly male. West Ross © Graham Sherwin

Fig. 230 Northern Emerald Dragonfly female. West Ross

Dragonfly Group 6

3) From above, thorax lacking bright-cream or yellow spot near the head (Fig 231a).
4) Thorax and abdominal segments 1 & 2 more conspicuously downy (Fig 231a).
5) From the front, head with frons lacking yellow markings (Fig 235a).
6) Wings with bases suffused heavily with orange (Fig 231a).
7) Wings with pterostigmata black rather than brown.

• **From male Brilliant Emerald Dragonfly**
1) From above, abdomen more conspicuously 'waisted' and club shaped (Fig 225).
2) From above, abdomen with claspers short and outwardly curving rather than long and straight (Fig 125a).
3) From above, abdomen with inferior anal appendage not usually visible (Fig 233a).

• **From female Brilliant Emerald Dragonfly**
1) From above, abdomen lacking yellow line at the junction of segments 2 & 3 (Fig 231a).
2) From above, abdomen lacking yellow

Fig. 231 Dorsal view of thorax, base of abdomen and wings
a. Downy Emerald Dragonfly, b. Northern Emerald Dragonfly © Graham Sherwin, c. Brilliant Emerald Dragonfly © Steve Cham, d. Brilliant Emerald Dragonfly © Fons Peels, DragonflyPix

Dragonfly Group 6

or cream spots on segment 3 (Fig 231a).
3) From above, abdomen with claspers short and thick rather than long and fine (Fig 233d).
4) From the side, abdomen with vulvar scale beneath segment 8 not usually visible rather than being long and protruding almost at right angles from the body.

Brilliant Emerald Dragonfly
Somatochlora metallica
(Vander Linden)

Male (Fig 227)
Female (Fig 228)
For species account see p 144.

Similar species
The Brilliant Emerald could easily be confused with the other two species in this group but differs in the following ways:

- **From both sexes of Downy Emerald Dragonfly**
1) Thorax bright apple-green rather than bronze-green (Fig 231d).
2) From above, thorax with fine, upwardly pointing, yellow or cream triangular marking emanating from bases of forewings (Figs 231c & d).
3) From above, thorax with bright-cream or yellow spot near the head (Figs 231c & d).
4) From above, thorax and abdominal segments 1 & 2 less conspicuously downy (Figs 231c & d).
5) From the front, frons with conspicuous elongated, yellow, 'U'- shaped marking which resembles a smiling face (Fig 235b).
6) Wings with bases not heavily suffused with orange (Figs 231c & d).
7) Wings with pterostigmata brown rather than black.

- **From male Downy Emerald Dragonfly**
1) From above, abdomen less conspicuously 'waisted' and club shaped (Fig 227).
2) From above, abdomen with superior claspers long and straight rather than short and outwardly curving (Fig 233b).
3) From above, abdomen with inferior anal appendage long, triangular and clearly visible from above (Fig 233b).

Fig. 232 Lateral view of thorax and abdomen a. Downy Emerald Dragonfly, b. Brilliant Emerald Dragonfly © Fons Peels, DragonflyPix, c. Northern Emerald Dragonfly

Dragonfly Group 6

- **From female Downy Emerald Dragonfly**

1) From above, abdomen with yellow or cream line at the junction of segments 2 & 3 (Figs 231c & d).
2) From above, abdomen of some individuals with a pair of yellow or cream patches or spots on segment 3 (Fig 231d).
3) From above, abdomen with claspers long and fine rather than short and thick (Fig 233e).
4) From the side, abdomen with vulvar scale beneath segment 8 long and protruding almost at right angles from the body (Fig 234b).

- **From both sexes of Northern Emerald Dragonfly**

1) Thorax and abdomen not tinged blue (Figs 227; 228).
2) Thorax robust rather than slender (Figs 227; 228).
3) From the front, frons with a yellow, elongated, 'U'- shaped band rather than a yellow spot at each side (Fig 235b).

- **From male Northern Emerald Dragonfly**

1) From above, abdomen thick-set rather than slender, particularly at segments 3 & 4 (Fig 227).
2) From above, abdomen with superior claspers long and straight rather than pincer-shaped (Fig 233b).
3) From above, abdomen with inferior anal appendage long and pointed rather than short and blunt (Fig 233b).

- **From female Northern Emerald Dragonfly**

1) From the side, abdomen with a series

Fig. 233 Dorsal view of claspers
a. Downy Emerald Dragonfly male, b. Brilliant Emerald Dragonfly male, c. Northern Emerald Dragonfly male © Graham Sherwin, d. Downy Emerald Dragonfly female, e. Brilliant Emerald Dragonfly female © Steve Cham f. Northern Emerald Dragonfly female

Dragonfly Group 6

of yellow spots on segment 2 rather than yellow or cream lines forming the outline shape of a Wellington boot (Fig 232b).
2) From the side abdomen with yellow or cream spot on upper half of segment 3 sometimes absent (Fig 232b).
3) From the side, abdomen with vulvar scale long and protruding almost at right angles from the body rather than short and protruding only at a slight angle (Fig 234b).
4) From above, abdomen with claspers long rather than short (Fig 233e).

Northern Emerald Dragonfly
Somatochlora arctica (Zetterstedt)

Male (Fig 229)
Female (Fig 230)
For species account see p 141.

Similar species
Superficially very similar to the other two species in this group but differs in the following ways:
• **From both sexes of Downy Emerald Dragonfly**
1) Thorax and abdomen tinged blue (Fig 229; 230).
2) Thorax slender rather than robust (Fig 231b).
3) Thorax and abdominal segments 1 & 2 less conspicuously downy (Fig 232c).
4) From above, thorax with yellow or cream triangle emanating from bases of forewings and yellow or cream spot near head (Fig 231b).
5) From the front, head with yellow spot at either side of frons (Fig 235c).
6) Wings with bases not suffused deeply with orange (Fig 231b).
7) Wings with pterostigmata brown rather than black.
• **From male Downy Emerald Dragonfly**
1) From above, abdomen more slender with long 'waist' and less conspicuously club shaped (Fig 229).
2) From above, superior claspers long and pincer shaped rather than short and outwardly curved (Fig 233c).
3) Flight patrols usually higher and not as close to the water's surface.

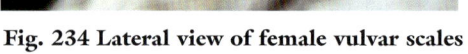

Fig. 234 Lateral view of female vulvar scales
a. Northern Emerald Dragonfly, b. Brilliant Emerald Dragonfly © Fons Peels, DragonflyPix

Dragonfly Group 6

Fig. 235 Facial markings a. Downy Emerald Dragonfly, b. Brilliant Emerald Dragonfly, c. Northern Emerald Dragonfly

- **From female Downy Emerald Dragonfly**

1) From above, abdomen with a pair of large, yellow or cream spots on segment 3 (Fig 231b).
2) From above, abdomen with claspers slender rather than stout (Fig 233f).
3) From the side, abdomen with linear markings on segment 2 that form the outline shape of a Wellington boot rather than a single large, yellow patch (Fig 232c).
4) From the side, abdomen with yellow spot on the upper half of segment 3 (Fig 232c).
5) From the side, abdomen with vulvar scale protruding at an angle from the body (Fig 234a).

- **From both sexes of Brilliant Emerald Dragonfly**

1) Thorax and abdomen tinged blue (Figs 229; 230).
2) Thorax slender rather than robust (Figs 229; 230).
3) Head with a yellow spot on each side of frons rather than an elongated, U-shaped bar (Fig 235c).

- **From male Brilliant Emerald Dragonfly**

1) Abdomen slender rather than thick-set, particularly at segments 3 & 4 (Fig 229).
2) From above, abdomen with superior claspers pincer shaped rather than straight (Fig 233c).
3) From above, abdomen with inferior anal appendage blunt and not obviously visible, rather than being long and pointed (Fig 233c).

- **From female Brilliant Emerald Dragonfly**

1) From the side, abdomen with yellow or cream linear markings forming the outline shape of a Wellington boot rather than a series of yellow spots (Fig 232c).
2) From the side, abdomen with vulvar scale beneath segment 8 short and protruding only at a slight angle from the body rather than long and protruding almost at right angles (Fig 234a).
3) From above, abdomen with pair of large, yellow spots on segment 3 always present (Fig 231b).
4) From above, abdomen with claspers shorter (Fig 233f).

Dragonfly Group 7
Medium sized, abdomen broad and dull brown, with cream lateral spots

Two species: Four-spotted Chaser adult male and form *praenubila*, Broad-bodied Chaser adult female

Four-spotted Chaser
Libellula quadrimaculata Linn.

Adult male (Fig 236)
Form *praenubila* (Fig 238b)
For species account see p 152.

Similar species
This species may appear superficially similar to the female Broad-bodied Chaser but differs in the following ways:
• **Adult male from female Broad-bodied Chaser**
1) From above, thorax lacking ante-humeral stripes (Fig 236).
2) From above, abdomen never tinged with blue (Fig 236).
3) Base of forewing lacking dark-brown or black patch (Fig 236).
4) Bases and leading margins of all wings suffused pale brown or chestnut (Fig 236).
5) All wings with a conspicuous dark spot at the nodes (Fig 238a).
6) Pterostigmata short and broad rather than long and slender (Fig 238a).
• **Form *praenubila* from female Broad-bodied Chaser**
Pterostigmata extended into large, smoky-black bands (Fig 238b).

Broad-bodied Chaser
Libellula depressa Linn.

Adult female (Fig 237)
For species account see p 148.

Similar species
The only species that resembles the female Broad-bodied Chaser is the Four-

Fig. 236 Four-spotted Chaser adult male Isle of Mull

Fig. 237 Broad-bodied Chaser adult female. South Hampshire

Dragonfly Group 7

Fig. 238 Wing markings a. Four-spotted Chaser, b. Four-spotted Chaser form *praenubila*, c. Broad-bodied Chaser female

spotted Chaser, from which it differs in the following ways:
1) From above, thorax with conspicuous antehumeral stripes (Fig 237).
2) From above, abdomen of older individuals tinged with blue (Fig 237).
3) Base of forewing with large, dark-brown or black patch (Fig 237).

4) Leading margins of all wings clear rather than suffused with brown (Fig 238c).
5) All wings lacking conspicuous dark spot at nodes (Fig 238c).
6) Pterostigmata long and slender rather than short and broad (Fig 238c).

Dragonfly Group 8
Medium-sized, abdomen blue

Four species: Scarce Chaser adult male, Broad-bodied Chaser adult male, Black-tailed Skimmer adult and sub-adult male, Keeled Skimmer adult and sub-adult male

Scarce Chaser
Libellula fulva Müller

Adult male (Fig 239)
For species account see p 150.

Similar species
The male Scarce Chaser is very similar in appearance to the adult males of the other three species in this group but differs in the following ways:

- **From male Broad-bodied Chaser**
1) From above, thorax lacking pale antehumeral stripes (Fig 245a).
2) From above, abdomen slender rather than broad (Fig 246a).
3) From above, abdomen without row of large, cream spots along each side (Fig 246a).
4) From above, abdomen with last three segments black rather than just the tip (Fig 246a).

Dragonfly Group 8

5) Forewing lacking dark patch at base (Fig 245a).
6) Hindwing with dark basal patch smaller and with yellow suffusion (Fig 245a).
7) Pterostigmata shorter (Fig 247a).
8) From above, eyes pale blue rather than dark brown (Fig 245a).
- **From male Black-tailed Skimmer**
1) From above, abdomen with dark 'tail' shorter (Fig 246a).
2) From above, abdomen with clear distinction between the blue segments and the dark 'tail' (Fig 246a).
3) Wings with dark areas at bases rather than being completely clear (Fig 245a).
4) Wings with costa clear rather than usually finely edged with pale yellow (Fig 247a).
5) From above, eyes pale blue rather than dark green (Fig 245a).
- **From male Keeled Skimmer**
1) Larger with more robust general build (Fig 239).
2) From above, thorax lacking antehumeral stripes (Fig 245a).
3) From above, abdomen broader (Fig 246a).
4) From above, abdomen with dark 'tail' rather than being entirely blue (Fig 246a).
5) Wings with dark areas at the bases rather than being completely clear (Fig 245a).
6) Pterostigmata black rather than amber (Fig 247a).
7) From above, eyes pale blue rather than dark green (Fig 245a).

Broad-bodied Chaser
Libellula depressa Linn.

Adult male (Fig 240)
For species account see p 148.

Similar species
The sub-adult forms of Black-tailed Simmer and Keeled Skimmer are very distinctively marked (Figs 246e & f) and should not be confused with the present species. However, the Broad-bodied Chaser is similar to the adult males of the other species in this group but differs in the following ways:
- **From male Scarce Chaser**
1) From above, thorax with conspicuous pale-blue antehumeral stripes (Fig 245b).
2) From above, abdomen broad rather than slender (Fig 246b).
3) From above, abdomen with a row of cream spots along each side (Fig 246b).
4) From above, abdomen with only terminal segment black rather than the last three (Fig 246b).
5) Forewing with large, dark patch at the base (Fig 245b).
6) Hindwing with dark basal patch larger and lacking yellow suffusion (Fig 245b).
7) Pterostigmata longer (Fig 247b).
8) From above, eyes dark brown rather than pale blue (Fig 245b).
- **From male Black-tailed Skimmer**
1) From above, thorax with conspicuous pale-blue antehumeral stripes (Fig 245b).
2) From above, abdomen broad rather than slender (Fig 246b).

Dragonfly Group 8

Fig. 239 Scarce Chaser adult male
West Sussex

Fig. 240 Broad-bodied Chaser adult male
South Hampshire

Fig. 241 Black-tailed Skimmer adult male
West Norfolk

Fig. 242 Black-tailed Skimmer sub-adult
male. East Norfolk

Fig. 243 Keeled Skimmer adult male
South Hampshire

Fig. 244 Keeled Skimmer sub-adult male
West Norfolk

Dragonfly Group 8

3) From above abdomen with a row of bold, cream spots along each side rather than plain blue (Fig 246b).
4) From above, abdomen with only terminal segment black rather than last four (Fig 246b).
5) Wings with dark basal patches (Fig 245b).
6) Forewing with costa clear rather than finely edged with yellow (Fig 247b).
7) Pterostigmata longer (Fig 247b).
8) From above, eyes dark brown rather than dark green (Fig 245b).

• **From male Keeled Skimmer**
1) Larger with more robust general build (Fig 240).
2) Thorax with antehumeral stripes broad and curved rather than slender and straight (Fig 245b).
3) Abdomen broad rather than slender (Fig 246b).
4) Abdomen with a row of bold, cream spots along each side (Fig 246b).
5) Abdomen with tip black rather than blue (Fig 246b).
6) Wings with dark basal patches (Fig 245b).

Fig. 245 Dorsal view of head, thorax and bases of wings
a. Scarce Chaser adult male, b. Broad-bodied Chaser adult male, c. Black-tailed Skimmer adult male, d. Keeled Skimmer adult male

Dragonfly Group 8

7) Pterostigmata black rather than amber (Fig 247b).
8) Eyes dark brown rather than dark green (Fig 245b).

Black-tailed Skimmer
Orthetrum cancellatum (Linn.)

Adult male (Fig 241)
Sub-adult male (Fig 242)
For species account see p 154.

Similar species
The adult male may be confused with any of the other species in this group and the sub-adult with that of Keeled Skimmer. However, the Black-tailed Skimmer differs in the following ways:

• **Adult male from adult male Scarce Chaser**
1) From above, abdomen with dark 'tail' longer (Fig 246c).
2) From above, abdomen with division between blue segments and the dark 'tail' not clear-cut (Fig 246c).
3) Wings clear and lacking dark basal patches (Fig 245c).
4) Wings with costa finely edged with yellow (Fig 247c).
5) From above, eyes dark green rather

Fig. 246 Dorsal view of abdomen a. Scarce Chaser adult male, b. Broad-bodied Chaser adult male, c. Black-tailed Skimmer adult male, d. Keeled Skimmer adult male, e. Black-tailed Skimmer sub-adult male, f. Keeled Skimmer sub-adult male

Dragonfly Group 8

than pale blue (Fig 245c).
• **Adult male from adult male Broad-bodied Chaser**
1) From above, thorax lacking antehumeral stripes (Fig 245c).
2) From above, abdomen slender rather than broad (Fig 246c).
3) From above, abdomen lacking bold, cream spots on each side (Fig 246c).
4) From above, abdomen with last four segments black rather than just the tip (Fig 246c).
5) Wings without dark basal patches (Fig 245c).
6) Wings with costa edged finely with yellow (Fig 247c).
7) Pterostigmata shorter (Fig 247c).
8) From above, eyes dark green rather than dark brown (Fig 245c).
• **Adult male from adult male Keeled Skimmer**
1) Generally more robust in build (Fig 241).
2) From above, thorax lacking antehumeral stripes (Fig 245c).
3) From above, abdomen broader (Fig 246c).
4) From above, abdomen with dark 'tail' rather than being entirely blue (Fig 246c).
5) Wings with costa edged finely with yellow (Fig 247c).
6) Pterostigmata dark grey or black rather than amber (Fig 247c).
• **Sub-adult male from sub-adult male Keeled Skimmer**
As above except for feature 4.
From above, abdomen with a series of large, yellow lateral patches (Fig 246e).

Keeled Skimmer
Orthetrum coerulescens
(Fabricius)

Adult male (Fig 243)
Sub-adult male (Fig 244)
For species account see p 157.

Similar species
The adult may be confused with that of any of the other three species in this group and the immature male with that of the Black-tailed Skimmer. However, the Keeled Skimmer differs in the following ways:
• **Adult male from adult male Scarce Chaser**
1) From above, thorax with antehumeral stripes (Fig 245d).
2) From above, abdomen much slimmer (Fig 246d).
3) From above, abdomen with blue rather than black tip (Fig 246d).
4) Wings lacking dark basal patches (Fig 245d).
5) Wings lacking dark, smoky patches at tips (Fig 247d).
6) Wings with pterostigmata amber-coloured rather dark brown (Fig 247d).
• **Adult male from adult male Broad-bodied Chaser**
1) From above, Broad-bodied Chaser much broader and stockier (Fig 243).
2) From above, eyes blue rather than dark brown (Fig 245d).
3) From above, abdomen lacking cream-coloured spots along margins (Fig 246d).
4) Wings lacking dark basal patches (Fig 245d).

Dragonfly Group 8

5) Wings with pterostigmata amber-coloured rather than dark brown (Fig 247d).
• **Adult male from adult male Black-tailed Skimmer**
1) From above, eyes blue rather than green (Fig 245d).
2) From above, thorax with well-defined antehumeral stripes (Fig 245d).
3) From above, abdomen lacking black tip (Fig 246d).
4) From above, abdomen lacking pale patches along the margins (Fig 246d).
5) Wings with pterostigmata amber-coloured rather than dark brown (Fig 247d).
6) Wings with costa not finely-edged with yellow (Fig 247d).
• **Sub-adult male from sub-adult male Black-tailed Skimmer**
1) Generally slimmer in build (Fig 244).
2) From above, thorax with well-defined antehumeral stripes (Fig. 244).
3) From above, abdomen lacking yellow lateral patches (Fig 246f).
4) Wings with pterostigmata amber-coloured rather than brown (Fig 244).

Fig. 247 Wing markings a. Scarce Chaser adult male, b. Broad-bodied Chaser adult male, c. Black-tailed Skimmer adult male, d. Keeled Skimmer adult male

Dragonfly Group 9
Medium-sized, abdomen orange or yellow

Four species: Scarce Chaser female and immature male, Broad-bodied Chaser female and immature male, Black-tailed Skimmer female and immature male, Four-spotted Chaser female

Scarce Chaser
Libellula fulva (Müller)

Female (Fig 248)
Immature male (Fig 249)
For species account see p 150.

Similar species
The brightness of the orange abdomen with its black, central band (Figs 248; 249) should preclude confusion with the other species in this group. The female can be separated from the immature male by the presence of large, grey, nebulous patches on the wing tips (Fig 256a).

Broad-bodied Chaser
Libellula depressa Linn.

Female (Fig 250)
Immature male (Fig 251)
For species account see p 148.

Similar species
The conspicuous antehumeral stripes on

Fig. 248 Scarce Chaser female. East Norfolk

Dragonfly Group 9

the thorax (Fig 255b), the broad abdomen with its lateral series of large, yellow spots (Fig 250; 251) and dark basal patches on all the wings (Fig 255b) should preclude confusion with the other species in this group.

Fig. 249 Scarce Chaser immature male
East Norfolk

Fig. 250 Broad-bodied Chaser female
West Norfolk

Fig. 251 Broad-bodied Chaser immature male. South Hampshire

Fig. 252 Black-tailed Skimmer female
West Norfolk

Fig. 253 Black-tailed Skimmer immature male. East Suffolk

Dragonfly Group 9

Black-tailed Skimmer
Orthetrum cancellatum (Linn.)

Female (Fig 252)
Immature male (Fig 253)
For species account see p 154.

Similar species
The striking, almost hornet-like, bright-yellow and black markings of the abdomen immediately separate this species from others in the group.

Fig. 254 **Four-spotted Chaser female East Norfolk**

Fig. 255 Dorsal view of thorax and bases of wings
a. Scarce Chaser, b. Broad-bodied Chaser, c. Black-tailed Skimmer, d. Four-spotted Chaser

Dragonfly Group 9

Four-spotted Chaser
Libellula quadrimaculata Linn.

Female (Fig 254)
For species account see p 152.

Similar species
The large, black spot at the nodes of each wing (Fig 256d) and the short, pointed abdomen (Fig 254) should immediately separate this species from others in the group.

Fig. 256 Wing markings a. Scarce Chaser, b. Broad-bodied Chaser, c. Black-tailed Skimmer, d. Four-spotted Chaser

Dragonfly Group 10
Small, abdomen red

Three species: Common Darter male and form *nigrescens* male (Highland Darter), Red-veined Darter male, Ruddy Darter male

Fig. 257 Common Darter male
West Norfolk

Fig. 258 Common Darter male form
nigrescens (Highland Darter). Isle of Mull

Common Darter
Sympetrum striolatum
(Charpentier)

Male (Fig 257)
Male form *nigrescens* (Highland Darter) (Fig 258)
For species account see p 165.

Similar species
All of the diagnostic features for the typical form of Common Darter are valid for form *nigrescens*. The latter only differs in being generally darker.
The Common Darter may easily be confused with the males of the other two species in this group but differs in the following ways:
• **From male Ruddy Darter**
1) From above, thorax sometimes with small antehumeral stripes (Fig 261a).
2) From the side, thorax with pale panels yellow rather than pale brown or pink (Fig 263a).
3) From above, abdomen of equal width throughout rather than being club shaped (Fig 257).
4) From above, abdomen brick-red rather than blood-red (Fig 257).
5) Upperside of abdomen with a varying number of pairs of pale-brown, square spots (Fig 257).
6) From the front, head with frons yellow rather than pink (Fig 263a).
7) From the front, frons edged only along the top, not on the sides, with black (Fig 263a).
8) From the side, upper half of eyes tinged brick-red rather than blood-red (Fig 263a).
9) Base of hindwing sometimes with small area of yellow suffusion.
10) Legs with yellowish stripes along

Dragonfly Group 10

their outer surfaces rather than being entirely black (Fig 263a).
• **From male Red-veined Darter**
1) From above, thorax sometimes with small antehumeral stripes (Fig 261a).
2) From the side, thorax with pale panels yellow rather than grey or pale blue (Fig 263a).
3) From above, abdomen with pairs of pale-brown, square spots rather than being plain (Fig 257).
4) From the front, head with frons yellow rather than pink (Fig 263a).
5) From the front, frons edged only along the top, not on the sides, with black (Fig 263a).
6) From the side, upper half of eyes tinged brick-red rather than blood-red (Fig 263a).
7) From the side, lower half of eyes pale brown rather than pale blue (Fig 263a).
8) Wings with veins along leading margin not tinged red (Fig 261a).
9) Pterostigmata dark and plain rather than pale with black edges (see Fig 262).
10) Base of hindwing without extensive yellow suffusion (Fig 261).

Fig. 259 Red-veined Darter male
France © Graham Sherwin

11) Legs with a conspicuous rather than fine, yellowish stripe along their outer surfaces (Fig 263a).

Red-veined Darter
Sympetrum fonscolombii (Sélys)

Male (Fig 259)
For species account see p 163.

Similar species
Easily confused with the other two species in this group but differs in the following ways:
• **From male Common Darter**
1) From above, thorax lacking antehumeral stripes (Fig 261b).
2) From the side, thorax with pale panels grey or pale blue rather than yellow (Fig 263b).
3) From above, abdomen lacking small, pale-brown, square spots (Fig 259).
4) From the front, head with frons tinged pink rather than yellow (Fig 263b).
5) From the front, frons edged with black on top and sides rather than the top only (Fig 263b).
6) From the side, upper half of eyes tinged blood-red rather than brick-red (Fig 263b).
7) From the side, lower half of eyes blue rather than pale brown (Fig 263b).
8) Wings with veins along leading margins tinged red (Fig 261b).
9) Pterostigmata pale yellow and strongly edged with black rather than being entirely dark (Fig 262).

Dragonfly Group 10

Fig. 260 Ruddy Darter male
West Norfolk

10) Hindwings with extensive yellow suffusion at bases (Fig 261b).
11) Legs with only a fine rather than conspicuous pale stripe along outer surfaces (Fig 263b).
- **From male Ruddy Darter**
1) From the side, thorax with pale panels grey, tinged with blue rather than pale brown and tinged pink (Fig 263b).
2) From above, abdomen of equal width throughout rather than being club shaped (Fig 259).

3) From the side, lower half of eyes pale blue rather than pale brown (Fig 263b).
4) Wings with veins along leading margins tinged red rather than being clear (Fig 261b).
5) Pterostigmata entirely dark rather than being pale yellow, strongly edged with black (Fig 262).
6) Hindwings with extensive yellow suffusion at bases (Fig 261b).

Ruddy Darter
Sympetrum sanguineum
(Müller)

Male (Fig 260)
For species account see p 168.

Similar species
Easily confused with the other two species in this group but differs in the following ways:
- **From male Common Darter**
1) From above, upperside of thorax lacking antehumeral stripes (Fig 261c).

Fig. 261 Dorsal view of male thorax, base of abdomen and wings
a. Common Darter, b. Red-veined Darter © Graham Sherwin, c. Ruddy Darter

Dragonfly Group 10

2) From the side, thorax with pale panels pink rather than yellow (Fig 263c).
3) From above, abdomen distinctly 'waisted' and club shaped (Fig 260).
4) From above, abdomen blood-red rather than brick-red (Fig 260).
5) From above, abdomen with pale-brown, square spots less extensive (Fig 260).
6) From the front, head with frons pink rather than yellow (Fig 263c).
7) From the front, frons edged black at the top and the sides rather than the top only (Fig 263c).
8) From the side, upper half of eyes tinged blood-red rather than brick-red (Fig 263c).
9) Wings usually lacking yellow, basal suffusion (Fig 261c).
10) Legs entirely black and without yellow stripe along outer surfaces (Fig 263c).

• **From male Red-veined Darter**
1) From the side, thorax with panels tinged pink rather than blue (Fig 263c).
2) From above, abdomen distinctly 'waisted' and club shaped (Fig 260).
3) From the side, lower half of eyes pale brown rather than blue (Fig 263c).
4) Wing-veins not tinged red (Fig 261c).
5) Pterostigmata entirely dark rather than pale yellow, edged with black (see Fig 262).
6) Hindwings lacking extensive yellow suffusion at bases (Fig 261c).
7) Legs lacking fine, yellowish stripe along their outer surfaces (Fig 263c).

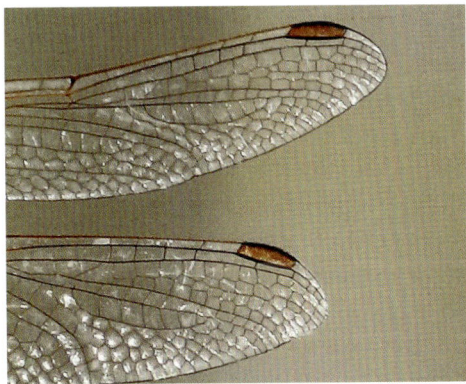

Fig. 262 Red-veined Darter pterostigmata
© Graham Sherwin

Fig. 263 Lateral view of male thorax, eyes and legs
a. Common Darter, b. Red-veined Darter © Graham Sherwin, c. Ruddy Darter

Dragonfly Group 11
Small, abdomen straw-coloured

Five species: Keeled Skimmer female and immature male, Common Darter male, female, immature male and f. *nigrescens* (Highland Darter), Red-veined Darter female and immature male, Black Darter female and immature male, Ruddy Darter female and immature male

Keeled Skimmer
Orthetrum coerulescens (Fab.)

Female (Fig 264)
Immature male (Fig 265)
For species account see p 157.

Similar species
Easily confused with other females and immature males in this group but differs in the following ways:

• **Female from female and immature male Common Darter**
1) From above, thorax with prominent, full-length rather than small antehumeral stripes (Fig 270a).
2) From the side, thorax lacking large, black-edged, yellow panels (Fig 273a).
3) From above, abdomen with a single black spot on each segment connected by a continuous black line (Fig 264).
4) From above, abdomen lacking isolated black markings on segments 8 & 9 (Fig 264).
5) From the side, abdomen with a single row of black, horizontal lines rather than two (Fig 273a).
6) From the side, eyes pale brown turning to pale blue with maturity rather than dark brown above and yellow below (Fig 273a).
7) From the front, head with frons not edged along the top with black (Fig 273a).
8) Legs pale brown rather than black with yellow stripes along their outer surfaces (Fig 273a).

**Fig. 264 Keeled Skimmer female
South Hampshire**

**Fig. 265 Keeled Skimmer immature male
West Norfolk**

Dragonfly Group 11

9) Wings with fore-margins suffused yellow rather than clear (Fig 270a).
10) Wings with costa white rather than black (Fig 276a).
11) Pterostigmata amber rather than dark brown (Fig 276a).

• **Immature male from female and immature male Common Darter**
As above plus:
1) From above, abdomen with a bold, dark, U-shaped marking on each segment and a continuous black, central line (Fig 265).
2) Eyes glassy, pale greenish-brown, turning blue with maturity rather than brown above and yellow below (Fig 273a).

• **Female from female Ruddy Darter**
1) From above, thorax with prominent, full-length antehumeral stripes (Fig 270a).
2) From above, abdomen with a single black spot on each segment connected by a continuous black line (Fig 264).
3) From above, abdomen lacking isolated black markings on segments 8 & 9 (Fig 264).
4) From the side, eyes pale brown turning to pale blue on maturity rather than brown on the upper half and greenish-yellow below (Fig 273a).
5) From the front, head with frons not edged with black along top and sides (Fig 273a).
6) Legs pale brown rather than black (Fig 273a).
7) Wings with costa white rather than black (Fig 276a).
8) Wings with fore-margin suffused yellow rather than clear (Fig 270a).
9) Hindwing lacking yellow/orange basal patch (Fig 270a).
10) Pterostigmata amber rather than dark brown (Fig 276a).

• **Immature male from female Ruddy Darter**
As above apart from features 4 & 8. Upperside of abdomen with a bold, dark, U-shaped marking on each segment and a continuous black, central line (Fig 265).

• **Female and immature male from immature male Ruddy Darter**
As above but also abdomen not distinctly club shaped (Figs 264; 265).

• **Female from female Black Darter**
1) From above, thorax with prominent, full-length antehumeral stripes (Fig 270a).
2) From above, thorax lacking distinctive downward-pointing, dark-brown or black, central triangle (Fig 270a).
3) From the side, thorax lacking bold, black and yellow markings (Fig 273a).
4) From above, abdomen with a single black spot on each segment connected by a continuous black line (Fig 264).
5) From above, abdomen lacking isolated black marking on segments 8 & 9 (Fig 264).
6) From the side, eyes pale brown turning to pale blue with maturity rather than brown on the upper half and green below (Fig 273a).
7) From the front, head with frons not edged with black along the top and sides (Fig 273a).
8) Legs pale brown rather than black

Dragonfly Group 11

(Fig 273a).
9) Wings with fore-margins suffused with yellow rather than clear (Fig 270a).
10) Wings with costa white rather than black (Fig 276a).
11) Pterostigmata amber rather than white or black (Fig 276a).

• **Immature male from female and immature male of Black Darter**
As above apart from points 4 & 9. Also: Upperside of abdomen with a bold, dark, U-shaped marking on each segment and a continuous black, central line (Fig 265).

• **Female and immature male from immature male Black Darter**
Abdomen not distinctly club shaped (Fig 264).

• **Female from female Red-veined Darter**
1) From above, thorax with antehumeral stripes slender and clear-cut rather than broad with diffuse edges (Fig 270a).
2) From the side, thorax lacking black-edged, pale grey-blue panels (Fig 273a).
3) From above, abdomen with a single black spot on each segment connected by a continuous black line (Fig 264).
4) From above, abdomen lacking isolated black markings on segments 2 & 3 (Fig 270a).
5) From above, abdomen lacking isolated black markings on segments 8 & 9 (Fig 264).
6) From the side, abdomen lacking series of pale rectangles edged strongly with black (Fig 273a).
7) From the side, eyes pale brown turning to pale blue with maturity rather than brown on the upper half and blue below (Fig 273a).
8) From the front, head with frons not edged with black along the top and sides (Fig 273a).
9) Legs pale brown rather than black with a pale line along their outer edges (Fig 273a).
10) Wings with fore-margins suffused with yellow rather than clear (Fig 270a).
11) Wings with costa white rather than yellow (Fig 270a).
12) Wing-veins in basal half black rather than yellow (Fig 270a).
13) Pterostigmata amber rather than pale yellow, edged strongly with black (Fig 276a).
14) Hindwings lacking yellow-suffused basal patch (Fig 270a).

• **Female from immature male Red-veined Darter**
As above apart from features 1 & 4.

• **Immature male from immature male Red-veined Darter**
As above (apart from features 3 & 10) but also:
Upperside of abdomen with a bold, dark, U-shaped marking on each segment and a continuous black, central line (Fig 265).

Common Darter
Sympetrum striolatum (Charpentier)

Female/immature male (Fig 266)
Female/immature male form *nigrescens* (Highland Darter) (Fig 267)
For species account see p 165.

Dragonfly Group 11

Similar species
The typical form of the Common Darter is easily confusable with the other species in this group but differs in the following ways:

• **From female Keeled Skimmer**
1) From above, thorax with antehumeral stripes short or absent rather than long and conspicuous (Fig 270b).
2) From the side, thorax with large, black-edged, yellow panels (Fig 273b).
3) From above, abdomen lacking large, black spot on each segment joined by a black, central line (Fig 266).
4) From the side, abdomen with two rows of black, horizontal lines rather than one (Fig 273b).
5) From the side, eyes brown above and yellow below rather than entirely pale brown or pale blue (Fig 273b).
6) From the front, head with frons edged along the top with black (Fig 273b).
7) Legs black with outer surfaces pale brown rather than completely brown (Fig 273b).
8) Wings with fore-margin clear rather than suffused yellow (Fig 270b).
9) Wings with costa black rather than white (Fig 270b).
10) Wings sometimes with very small, yellow basal patch (Fig 266).
11) Pterostigmata dark brown rather than amber (Fig 276b).

• **From immature male Keeled Skimmer**
As above plus:
From above, abdomen lacking bold, dark, U-shaped marking on each segment and a continuous black, central line (Fig 266).

• **From female Ruddy Darter**
1) From above, thorax sometimes with small antehumeral stripes (Fig 270b).
2) From the side, thorax with yellow rather than pale-brown, black-edged panels (Fig 273b).
3) From the front, head with frons edged with black along top only rather than the top and sides (Fig 273b).

Fig. 266 Common Darter immature male West Norfolk

Fig. 267 Common Darter female form *nigrescens* (Highland Darter). Isle of Mull

Dragonfly Group 11

4) Legs with yellow stripe along outer surfaces rather than being entirely black (Fig 273b).
5) Wings with less, if any, yellow suffusion at their bases (Fig 270b).
- **From immature male Ruddy Darter**
As above but also:
Abdomen not distinctly club shaped (Fig 266).

- **From female Black Darter**
1) From above, thorax usually with small antehumeral stripes (Fig 270b).
2) From above, thorax lacking large, dark-brown, downward-pointing, central triangle (Fig 270b).
3) From the side, thorax with two broad rather than narrow, yellow panels (Fig 273b).
4) From the side, thorax lacking black, central panel containing three isolated yellow spots (Fig 273b).
5) From above, abdomen with segments 8 & 9 lacking four conspicuous yellow spots on dark background (Fig 266).
6) From the front, head with frons not edged at the sides with black (Fig 273b).
7) Legs with a pale line along their edges rather than being entirely black (Fig 273b).
8) Pterostigmata brown rather than white or black (Fig 276b).
- **From immature male Black Darter**
As above but also:
1) Abdomen not distinctly club shaped (Fig 266).
2) From above, abdomen lacking black edges to segments 5–7 (Fig 266).

- **From female Red-veined Darter**
1) From above, when present, thorax with antehumeral stripes short and inconspicuous rather than long and bold (Fig 270b).
2) From the side, thorax with pale panels yellow rather than grey or pale blue (Fig 273b).
3) From above, base of abdomen lacking small, black, central markings on segments 2 & 3 (Fig 270b).
4) From the side, abdomen lacking series of pale rectangles edged strongly with black (Fig 273b).
5) From the front, head with frons not edged along sides with black (Fig 273b).
6) Wings with costa black rather than yellow (Fig 276b).
7) Wings with veins in basal half black rather than yellow (Fig 270b).
8) Hindwings lacking extensive yellow suffusion at bases (Fig 270b).
9) Pterostigmata dark brown rather than yellow edged with black (Fig 270b).
- **From immature male Red-veined Darter**
As above apart from features 1 & 3.

The Scottish Highland form *nigrescens* may be seen in company with Keeled Skimmer and Black Darter, from which it differs in the following ways:
- **Female from female and immature male Keeled Skimmer**
1) From above, thorax with antehumeral stripes short rather than long (Fig 267).
2) From the side, thorax with bold, black and yellow markings (Fig 267).
3) Legs black with a pale line along their outer surfaces rather than plain brown (Fig 267).

Dragonfly Group 11

4) Pterostigmata dark brown or black rather than amber (Fig 276b).
• **Female from female Black Darter**
1) From above, thorax with antehumeral stripes present (Fig 267).
2) From above, thorax lacking large, brown or black, downward-pointing, central, triangular marking (Fig 270b).
3) Legs with pale line along their outer surfaces rather than being entirely black (Fig 267).

Red-veined Darter
Sympetrum fonscolombii
(Sélys)

Female (Fig 268)
Immature male (Fig 269)
For species account see p 163.

Similar species
Easily confused with other members of the group but differs in the following ways:
• **Female from female Keeled Skimmer**
1) Upperside of thorax with antehumeral stripes broad with diffuse edges rather than slender and clear-cut (Fig 270e).
2) Side of thorax with two pale-grey or pale-blue, black-edged panels (Fig 273e).
3) Upperside of abdomen lacking central series of black spots connected by a black line (Fig 268).
4) Upperside of abdomen with isolated black markings on segments 2 & 3 (Fig 270e).
5) Upperside of abdomen with isolated black markings on segments 8 & 9 (Fig 268).

6) Eyes brown above and blue below, rather than being entirely pale brown or pale blue (Fig 273e).
7) Frons edged with black along the top and sides (Fig 273e).
8) Legs black with a pale line along their outer surfaces rather than being entirely pale brown (Fig 273e).
9) Outer half of wings clear with fore-margins not suffused yellow (Fig 268).

**Fig. 268 Red-veined Darter female
South Hampshire** © Steve Cham

**Fig. 269 Red-veined Darter immature male
Cyprus**

Dragonfly Group 11

10) Wings with costa yellow rather than white (Fig 276d).
11) Wing-veins in basal half yellow rather than black (Fig 270e).
12) Pterostigmata pale yellow, edged strongly with black rather than wholly amber (Fig 276d).
13) Hindwings with yellow-suffused basal patch rather than being clear (Fig 270e).

• **Immature male from female Keeled Skimmer**
As above apart from features 1 & 4.

• **Female and immature male from immature male Keeled Skimmer**
Upperside of abdomen lacking large, bold markings (Figs 268; 269).

• **Female from female and immature male of Common Darter**
1) From above, thorax with antehumeral stripes long and bold rather than short or absent (Fig 270e).
2) From the side, thorax with pale panels grey or pale blue rather than yellow (Fig 273e).
3) From above, abdomen with small, black, central markings on segments 2 & 3 (Fig 270e).
4) From the side, abdomen with a series of pale rectangles edged strongly with black (Fig 269).
5) From the front, head with frons edged with black along top and sides rather than the top only (Fig 273e).
6) Costa yellow rather than black (Fig 276d).
7) Wings with veins yellow in basal half rather than black (Fig 270e).
8) Hindwings with more extensive yellow suffusion at bases (Fig 270e).
9) Pterostigmata yellow, edged with black rather than wholly dark brown (Fig

Fig. 270 Dorsal view of thorax and bases of wings a. Keeled Skimmer, b. Common Darter c. Ruddy Darter, d. Black Darter, e. Red-veined Darter

Dragonfly Group 11

276d).
- **Immature male from female and immature male of Common Darter**
As above apart from features 1 & 3.

- **Female from female Ruddy Darter**
1) Upperside of thorax with antehumeral stripes present (Fig 270e).
2) Side of thorax with pale panels grey or pale blue rather than yellowish-brown (Fig 273e).
3) Upperside of abdomen with small, isolated, black, central markings on segments 2 & 3 (Fig 270e).
4) Side of abdomen with series of pale rectangles edged strongly with black (Fig 269).
5) Eyes blue on lower half rather than yellowish-green (Fig 273e).
6) Wings with costa yellow rather than black (Fig 276d).
7) Wing-veins in basal half of forewing yellow rather than black (Fig 268).
8) Hindwing with yellow basal patch larger (Fig 270e).
9) Pterostigmata pale yellow, edged strongly with black rather than being wholly dark brown (Fig 276d).
10) Legs with pale line along their outer surfaces rather than being entirely black (Fig 273e).
- **Immature male from female Ruddy Darter**
As above apart from feature 3.
- **Immature male from immature male Ruddy Darter**
As for female (apart from feature 3) but also:
Abdomen not club shaped (see Fig 275).

- **Female from female Black Darter**
1) Upperside of thorax with antehumeral stripes present (Fig 270e).
2) Upperside of thorax without dark-brown, downward-pointing, triangular marking (Fig 270e).
3) Side of thorax without bold pattern of black and yellow bars and spots (Fig 273e).
4) Upperside of abdomen with small, black, central markings on segments 2 & 3 (Fig 270e).
5) Upperside of abdomen lacking four yellow spots in a brown background on segments 8 & 9 (Fig 268).
6) Side of abdomen with a series of pale rectangles edged with black (Fig 273e).
7) Frons finely, rather than heavily, edged with black along top and sides (Fig 273e).
8) Lower half of eyes blue rather than yellowish-green (Fig 273e).
9) Wings with veins in basal half of forewing yellow rather than black (Fig 270e).
10) Costa yellow rather than black (Fig 276d).
11) Hindwing bases suffused yellow rather than clear (Fig 270e).
12) Pterostigmata yellow, edged strongly with black rather than wholly white or black (Fig 276d).
- **Immature male from female Black Darter**
As above apart from feature 4.
- **Immature male from female and immature male of Black Darter**
As above (apart from feature 4) but also:
Abdomen not distinctly club shaped (see Fig 272).

Dragonfly Group 11

Fig. 271 Black Darter female
South Hampshire

Fig. 272 Black Darter immature male
East Norfolk

Black Darter
Sympetrum danae (Sulzer)

Female (Fig 271)
Immature male (Fig 272)
For species account see p 160.

Similar species
Easily confused with the other species in this group but differs in the following ways:
- **Female from female Keeled Skimmer**
1) From above, thorax lacking bold antehumeral stripes (Fig 270d).
2) From above, thorax with prominent, downward-pointing, dark-brown, triangular marking (Fig 270d).
3) From the side, thorax with bold, black and yellow markings (Fig 273d).
4) From above, abdomen lacking black, central line with a black spot on each segment (Fig 271).
5) From above, abdomen with four pale spots surrounded by dark brown on segments 8 & 9 (Fig 271).
6) From the side, eyes brown above and yellowish-green below rather than being wholly pale brown or pale blue (Fig 273d).
7) From the front, head with frons edged with black along the top and sides (Fig 273d).
8) Legs black rather than pale brown (Fig 273d).
9) Wings with fore-margins clear rather than suffused with yellow (Fig 270d).
10) Pterostigmata black or white rather than amber (Figs 276e & f).
- **Female from immature male Keeled Skimmer**
As above apart from features 4 & 9. Also: Upperside of abdomen lacking bold markings (Fig 271).
- **Immature male from female and immature male of Keeled Skimmer**
As for female but also:
Abdomen distinctly club shaped (Fig 272).

- **Female from female and immature male Common Darter**

Dragonfly Group 11

1) From above, thorax always lacking antehumeral stripes (Fig 270d).
2) From above, thorax with large, downward-pointing, dark-brown triangle (Fig 270d).
3) From the side, thorax with bold pattern of narrow rather than broad, yellow and black bars (Fig 273d).
4) From the side, thorax with a row of three isolated yellow spots in black, central band (Fig 273d).
5) From above, abdomen with four yellow spots surrounded by dark brown on segments 8 & 9 (Fig 271).
6) From the front, head with frons edged along top and sides, rather than top only, with black (Fig 273d).
7) Legs entirely black rather than having a pale-yellow line along the outer surfaces (Fig 273d).
8) Pterostigmata white or black rather than brown (Figs 276e & f).

• **Immature male from female and immature male Common Darter**
As above but also:

From above, abdomen distinctly club shaped with black edges to upperside of segments 5–7 (Fig 272).

• **Female from female Ruddy Darter**
1) From above, thorax with large, dark-brown, downward-pointing, triangular marking (Fig 270d).
2) From the side, thorax with bold pattern of yellow and black bars and spots (Fig 273d).
3) From above, abdomen lacking four yellow spots on brown background on segments 8 & 9 (Fig 271).
4) Bases of wings with smaller patches of yellow suffusion (Fig 270d).
5) Pterostigmata white or black rather than pale grey or dark brown (Figs 276e

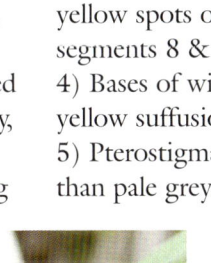

Fig. 273 Lateral view of thorax, eyes and legs a. Keeled Skimmer, b. Common Darter, c. Ruddy Darter, d. Black Darter, e. Red-veined Darter

Dragonfly Group 11

& f).
• **Female from immature male Ruddy Darter**
As above but also:
Abdomen not distinctly club shaped (Fig 271).
• **Immature male from immature male Ruddy Darter**
As for female but also:
From above, abdominal segments 5–7 with bold, black edges (Fig 272).

• **Female from female Red-veined Darter**
1) From above, thorax lacking pale antehumeral stripes (Fig 270d).
2) From above, thorax with large, dark-brown, downward-pointing, triangular marking (Fig 270d).
3) From the side, thorax with bold pattern of black and yellow bars and spots (Fig 273d).
4) From above, abdomen lacking small, black, central markings on segments 2 & 3 (Fig 270d).
5) From above, abdomen with four yellow spots on a brown background on segments 8 & 9 (Fig 271).
6) From the front, head with frons heavily rather than finely, edged with black along top and sides (Fig 273d).
7) From the side, lower half of eyes yellowish-green rather than blue (Fig 273d).
8) Wings with veins in basal half black rather than yellow (Fig 270d).
9) Costa black rather than yellow (Fig 270d).
10) Hindwing bases lacking yellow suffusion (Fig 270d).
11) Pterostigmata white or black rather than pale yellow, edged strongly with black (Figs 276e & f).
• **Female from immature male Red-veined Darter**
As above apart from feature 4.
• **Immature male from female and immature male of Red-veined Darter**
As above (apart from feature 4) but also: Abdomen distinctly club shaped with black edges to upperside of segments 5–7 (Fig 272).

Ruddy Darter
Sympetrum sanguineum (Müller)

Female (Fig 274)
Immature male (Fig 275)
For species account see p 168.

Similar species
Easily confused with the other members of this group but differs in the following ways:
• **Female from female Keeled Skimmer**
1) From above, thorax lacking antehumeral stripes (Fig 270c).
2) From above, abdomen lacking black spot and conspicuous black, central line on each segment (Fig 274).
3) From above, abdomen with an isolated black spot or bar on segments 8 & 9 (Fig 274).
4) From the side, eyes brown above and greenish-yellow below rather than wholly pale greenish-brown or pale blue (Fig 273c).
5) From the front, head with frons edged

along top and sides with black (Fig 273c).
6) Legs black rather than pale brown (Fig 273c).
7) Wings with costa black rather than white (Fig 270c).
8) Wings with fore-margins clear rather than suffused yellow (Fig 270c).
9) Hindwings with yellow-orange suffusion at the extreme base rather than clear (Fig 270c).
10) Pterostigmata grey or dark brown rather than amber (Fig 276c).

• **Female from immature male Keeled Skimmer**
As above apart from features 4 & 9. Also:
From above, abdomen lacking bold markings (Figs 274).

• **Immature male from female and immature male Keeled Skimmer**
As above but abdomen distinctly club shaped (Fig 275).

• **Female from female Common Darter**
1) From above, thorax lacking antehumeral stripes (Fig 270c).
2) From the side, thorax lacking bright-yellow, black-edged panels (Fig 273c).
3) From the front, head with frons edged with black along top and sides rather than the top only (Fig 273c).
4) Legs entirely black without pale-yellow stripe along outer surfaces (Fig 273c).
5) Wings with yellow-orange suffusion at bases rather than clear (Fig 276c).

• **Immature male from female and immature male Common Darter**
As above but abdomen distinctly club shaped (Figs 275).

• **Female from female Black Darter**
1) From above, thorax lacking large, dark-brown, downward-pointing, tri-angular marking (Fig 270c).
2) From the side, thorax lacking bold, bright pattern of yellow and black bars and spots (Fig 273c).
3) From above, abdomen lacking the yellow spots surrounded by black on segments 8 & 9 (Fig 274).

Fig. 274 Ruddy Darter female
West Norfolk

Fig. 275 Ruddy Darter immature male
West Norfolk

Dragonfly Group 11

4) Bases of wings with larger patches of yellow-orange suffusion (Fig 270c).
5) Pterostigmata pale grey or dark brown rather than white or black (Fig 276c).
- **Immature male from female Black Darter**
As above but also:
Abdomen clearly club shaped (Fig 275).
- **Immature male from immature male Black Darter**
As for female but also:
From above, abdominal segments 5–7 lacking bold, black edges (Fig 275).

- **Female from female Red-veined Darter**
1) From above, thorax lacking bold antehumeral stripes (Fig 270c).
2) From the side, thorax with pale panels yellowish-brown rather than grey or pale blue (Fig 273c).
3) From above, abdomen lacking small, black, central markings on segments 2 & 3 (Fig 270c).
4) From the side, abdomen lacking series of pale rectangles strongly edged with black (Fig 273c).
5) From the side, head with frons yellow rather than tinged pale blue (Fig 273c).
6) From the side, eyes yellowish-green on lower half rather than blue (Fig 273c).
7) Wings with costa black rather than yellow (Fig 276c).

Fig. 276 Pterostigmata a. Keeled Skimmer, b. Common Darter, c. Ruddy Darter, d. Red-veined Darter, e. Black Darter adult, f. Black Darter immature

Dragonfly Group 11

8) Wing-veins in basal half black rather than yellow (Fig 270c).
9) Bases of wings with area of yellow suffusion smaller (Fig 274).
10) Pterostigmata dark brown rather than pale yellow edged strongly with black (Fig 276c).
11) Legs entirely black rather with fine, pale line along outer surfaces (Fig 273c).

- **Female from immature male Red-veined Darter**
As above apart from feature 3.
- **Immature male from immature male Red-veined Darter**
As for female (apart from feature 3) but also:
Abdomen distinctly club shaped (Fig 275).

Dragonfly Group 12
Small, abdomen black and red
One species: White-faced Darter adult male

White-faced Darter
Leucorrhinia dubia (Vander Linden)

Adult male (Figs 277; 278)
For species account see p 146.

Similar species
None. The black and red body markings and conspicuous white 'face' preclude confusion with any other species found in Britain and Ireland.

Fig. 277 White-faced Darter adult male Shropshire © Keith Mallet

Fig. 278 White-faced Darter adult male face. Shropshire © Keith Mallet

Dragonfly Group 13
Small, abdomen black and yellow
Two species: White-faced Darter female and immature male, Black Darter adult male

White-faced Darter
Leucorrhinia dubia
(Vander Linden)

Immature male (Fig 279)
Female (Fig 280)
For species account see p 146.

Similar species
Often found in the company of Black Darter, but the present species does not have a club-shaped abdomen (Figs 279; 280). This, and the conspicuous yellow 'face' (Fig 284), is diagnostic of the White-faced Darter.

Fig. 280 White-faced Darter female Shropshire

Black Darter
Sympetrum danae (Sulzer)

Adult male (Fig 281)
For species account see p 160.

Similar species
Sometimes found in the company of White-faced Darter, but the club-shaped abdomen (Fig 281) precludes confusion with that species.

Fig. 279 White-faced Darter immature male. Shropshire

Fig. 281 Black Darter adult male South Hampshire

Dragonfly Group 13

Fig. 282 Dorsal view of thorax and base of abdomen and wings a. Black Darter adult male, b. White-faced Darter immature male, c. White-faced Darter female

Fig. 283 Lateral view of thorax a. Black Darter, b. White-faced Darter immature male, c. White-faced Darter female © Peter Vandrome

Fig. 284 White-faced Darter immature male face

Dragonfly Group 14
Small to medium-sized, abdomen dark smoky-brown

Four species: Scarce Chaser over-mature, Black-tailed Skimmer over-mature, Keeled Skimmer over-mature, Common Darter over-mature

Scarce Chaser
Libellula fulva Müller

Over-mature individual (Fig 285)
For species account see p 150.

Similar species
Superficially similar to the other three species in this group but differs in the following ways:
1) All wings with distinctive black tip (Fig 285).
2) Hindwings with dark basal area (Fig 285).

Black-tailed Skimmer
Orthetrum cancellatum (Linn.)

Over-mature individual (Fig 286)
For species account see p 154.

Similar species
Superficially similar to the other species in this group but differs in the following way:
From above, abdomen with conspicuous pale lateral bar aside each segment (Fig 286).

Fig. 285 Scarce Chaser over-mature West Sussex

Fig. 286 Black-tailed Skimmer over-mature. South Hampshire

Dragonfly Group 14

Keeled Skimmer
Orthetrum coerulescens
(Fabricius)

Over-mature individual (Fig 287)
For species account see p 157.

Similar species
Superficially similar to the other species in this group but differs in the following way:
Pterostigmata amber in colour (Fig 287).

Common Darter
Sympetrum striolatum
(Charpentier)

Over-mature individual (Fig 288)
For species account see p 165.

Similar species
The more slender abdomen combined with dark-brown or black pterostigmata should preclude confusion with the other species in this group (Fig 288).

Fig. 287 Keeled Skimmer over-mature South Hampshire

Fig. 288 Common Darter over-mature West Norfolk

CHAPTER FIVE
Annual Visitors and probable Colonists

The following species are now seen in Britain annually. Dainty Damselfly and Southern Migrant Hawker maintain at least transient populations and may become permanently established residents. Southern Emerald Damselfly and Lesser and Vagrant Emperor have at least attempted to breed here. The following account details their history, occurrence and distinguishing features.

Southern Emerald Damselfly
Lestes barbarus (Fab.)

History
First recorded in Britain in 2002 at Winterton Dunes, East Norfolk. Since its discovery, small numbers of the species have been reported almost annually in several eastern and south-eastern counties of England with a single record from North Somerset in 2006 (Cham *et al.*, 2014). Almost all of the records have come from coastal hinterland localities. There have been several attempts at breeding with presumed modest success at one locality in East Kent. It appears, however, that the species is struggling to establish a strong foothold there.

Flight period
Most records have occurred between late June and early September.

Habitat preferences
Ephemeral ponds, scrapes and ditches in coastal hinterland localities. The species is tolerant of brackish situations.

Distinguishing features
Superficially similar to the species discussed in Damselfly Group 4 (p 70) but differs most significantly in having the pterostigmata conspicuously bicoloured (Fig 289).

Dainty Damselfly
Coenagrion scitulum (Ramb.)

History
First discovered in 1946 at a site in South Essex where a strong colony existed until coastal flooding destroyed

Fig. 289 Southern Emerald Damselfly East Norfolk

Annual Visitors

the species' habitat in 1953. It was thereafter declared extinct as a British damselfly. An unsuccessful reintroduction attempt took place in 1966. A few individuals, including exuviae, have been found each year since 2010 in East Kent. It is now known from three, or possibly four, sites in the county, but populations at all these sites are small (A. Parr, pers. com.). This suggests that the Dainty Damselfly is attempting to re-colonise part of southern England but is struggling to maintain a foothold.

Fig. 290 Dainty Damselfly. East Kent
© Gary Thoburn

Flight period
Usually late May to early July.

Habitat preferences
Open and sunny dykes, ditches and ponds with rich aquatic vegetation into which the eggs are laid.

Distinguishing features
Similar in general appearance to the species discussed in Damselfly Group 9 (p 85) but is characterised by the long, amber pterostigmata. These are very much more elongate (twice as long as they are wide) rather than approximately square.

Souther Migrant Hawker
Aeshna affinis Vander Lind.

History
First recorded in 1952 in East Kent. There was a large influx of the species in 2010 which has led to at least transient

Fig. 291 Southern Migrant Hawker male
South Essex © Simon Lammin

Fig. 292 Southern Migrant Hawker female
South Essex © Simon Lammin

Annual Visitors

colonisation of few sites in South Essex and East Kent. Nationally, the number of annual sightings has increased considerably, and the species has been recorded regularly south-east of a line approximately between West Cornwall in the west and West Norfolk in the east. There are further records from North Lincolnshire, South-east Yorkshire and the area straddling South Lancashire and Cheshire (A. Parr, pers. com.).

Flight Period
In England this usually begins in June, and sightings have been reported until early September.

Fig. 293 Lateral view of male thorax
a. Southern Migrant Hawker © Steve Lammin **b. Migrant Hawker**

Fig. 294 Female thorax and base of abdomen
a. Southern Migrant Hawker © Steve Lammin, **b. Migrant Hawker**

Annual Visitors

Habitat preferences
Drainage channels and ponds in marshland. In England these are often prone to drying out during the summer.

Distinguishing features
• Male
Superficially resembles those species in Dragonfly Group 3. It is most similar to the male Migrant Hawker but differs in the following ways:
1) From the side, thorax plain yellow with fine, black lines at the intersegmental divisions.
2) From the side, thorax without broad, conspicuous diagonal bands (Fig 293a).
3) From above, abdomen black and bright blue. There are no yellow markings (Fig 291).
• Female
Superficially resembles some of those species in Dragonfly Group 4. It is most similar to the female Migrant Hawker but differs in the following ways:
1) From the side, thorax as described for the male (Fig 294a).
2) From above, segment 1 with yellow, triangular spot above the 'golf-tee' marking that is absent in Migrant Hawker (Fig 294a).

Lesser Emperor Dragonfly
Anax parthenope (Sélys)

History
First discovered at a site in Gloucestershire in 1996. Since then sightings have increased on an almost annual level and the species has now been recorded each year, though erratically, over most of England, southern Wales and Eastern Ireland. There is a single record from the Orkney Isles. Ovipositing has been reported on several occasions and the discovery of exuviae has proved breeding on several occasions.

Flight period
Sightings have occurred between May and September with an apparent peak during July and August.

Habitat preferences
This is very much a species that could occupy many types of lowland habitats.

Distinguishing features
The distinctive blue 'saddle' on segments 1 and 2 of the male, and most females, precludes confusion with any of our resident species. The dragonfly with which it is most likely to be confused is the Vagrant Emperor (see following species) from which it differs in the

Fig. 295 Lesser Emperor Dragonfly female. France © Graham Sherwin

Annual Visitors

following ways:
1) From above, eyes green rather than brown.
2) From above, base of abdomen with conspicuous yellow ring on segment 2.
3) Wings clear and lacking orange-yellow costa and basal veins.

Vagrant Emperor Dragonfly
Anax ephippiger (Burm.)

Fig. 296 Vagrant Emperor Dragonfly
© Paul Winter

History
First recorded in the British Isles in 1903 at Devonport in South Devon. Since the second specimen was discovered in 1968 the species has undergone a long history of 'feast or famine' when, during many years, it has been seen either scarcely or not at all, or occasionally (e.g. 2011 and 2013) in large numbers (Asher *et al.*, 2014). The most dramatic influx was during 2019 when hundreds were noted throughout the British Isles. In England, Wales and Scotland there are records from as far south as the Isles of Scilly and northwards to Orkney and even Shetland. However, in Ireland it has only been found in the counties of the extreme south and east. It is native to Afro-tropical regions and has recently colonised parts of the Mediterranean area.

Flight period
With the possible exception of December, the Vagrant Emperor has been recorded in the British Isles during every month of the year. With what is often a scarce visitor, it is difficult to identify its most likely time of arrival. However, the data presented by Asher *et al.*, 2014) and those gleaned from the pager service *Rare Bird Alert* suggest strongly that in the exceptional year 2019 there were three peaks of appearance. These were during February, June and October. It must be appreciated that these dates are merely guidelines and, as stated earlier, the species may be encountered at any time.

Habitat preferences
As a powerful migrant, the Vagrant Emperor may be seen anywhere and in many habitat types. British and Irish records show a distinct bias towards coastal areas and their hinterland, but inland lakes and reservoirs have also proved attractive.

Distinguishing features
Most easily confused with the previous species. See text and Fig. 296.

CHAPTER SIX
A Review of scarce or assisted Visitors and extinct Residents

Below is given a short account of those species that are now considered to be extinct as British residents and those that appear here only as very scarce or perhaps assisted vagrants. It is not felt to be within the remit of this book to illustrate all of these, but images can be found in, for example, Dijkstra & Lewington (2006), Brooks & Lewington (2004), Smallshire & Swash (2014) and, of course, the Internet.

Scarce or assisted Visitors

Winter Damselfly
Sympecma fusca (Vander Lind.)
A single record from West Glamorgan in 2008 (Parr, 2009). Native to most of Europe and parts of North Africa.

Marsh Bluetail
Ischnura senegalensis (Ramb.)
Although most 'exotic' species encountered in Britain are found indoors and usually associated with, for example, aquaria and their imported plants, a single specimen of the Marsh Bluetail was found at a garden pond in Kent in 2010 (Parr, 2010). This species is found naturally from Africa eastwards to Japan.

Green Darner
Anax junius Drury
Only six records in 1998 from West Cornwall and the Isles of Scilly (Pellow, 1999 and Corbet, 2000). Native to North America.

Yellow-legged Clubtail
Stylurus flavipes (Charp.)
A single record from East Sussex in 1818 (Lucas, 1900). Native to Europe east of France along major river courses.

Yellow-spotted Emerald
Somatochlora flavomaculata (Vander Linden)
A single individual was seen at Carlton Marshes, East Suffolk on 2nd July 2018. As the species has not previously been illustrated in a book, the following images are included here with the kind permission of the discoverer Andrew Easton.

Fig. 297 Yellow-spotted Emerald East Suffolk © Andrew Easton

Scarce or assisted Visitors and extinct Residents

Fig. 298 Yellow-spotted Emerald East Suffolk © Andrew Easton

Scarlet Darter
Crocothemis erythraea (Brullé)
Only seven records since its first sighting in West Cornwall in 1995 (Jones, 1996). Native to southern and eastern Europe.

Large White-faced Darter
Leucorrhinia pectoralis (Charp.)
First recorded in East Kent in 1859 (Lucas, 1900). There have since been two further records, both from East Suffolk, in 2012. Native at scattered localities in Europe east of France.

Wandering Glider
Pantala flavescens (Fab.)
Only three records: Horning, East Norfolk, 1823, Bolton, Lancashire, 1951 and "Kent", 1989 (Cham *et al.*, 2014). Of global distribution but scarce in Europe.

Yellow-winged Darter
Sympetrum flaveolum (Linn.)
Very scarce, occasionally irruptive, visitor. During large immigrations the species has been recorded at many localities in England and Wales with a few records from Scotland, Ireland and the Isles of Scilly. Breeding has been proven very occasionally but the 'colonies' are short-lived. In western and central Europe, the distribution is somewhat fragmented, being more widespread in the east.

Banded Darter
Sympetrum pedemontanum ([Müll.])
Only one confirmed record from Breconshire in 1995 (Cham *et al.*, 2014). Scarce and localised in western Europe but becoming more common in the east. Cham *et al.* (2014) state that the species is spreading gradually westwards.

Vagrant Darter
Sympetrum vulgatum (Linn.)
An extremely scarce vagrant that has been seen on only 25 occasions. Many of these records are historic. It has been reported mainly from southern and south-eastern England but never from Scotland, Wales or Ireland. Native in much of northern and eastern Europe.

Scarce or assisted Visitors and extinct Residents

Extinct Residents

Norfolk Damselfly
Coenagrion armatum (Charp.)
Found formerly in an area centred on Sutton, Ranworth and Hickling Broads in East Norfolk between 1919 and 1958 (Brownett, 2005). In mainland Europe the species inhabits shallow (and therefore presumably relatively warm) water with lush surface and emergent vegetation. The great increase in the recreational use of the Broads since the 1950's may have had an ultimately disastrous effect on the species' habitat requirements.

Orange-spotted Emerald
Oxygastra curtisii (Dale)
Always of extremely restricted distribution in Britain where the only known breeding site was centred on a short stretch of the River Moors near Hurn in Dorset. It was recorded here between 1820 and 1963. Other specimens were collected from nearby Pokesdown between 1878 and 1905. The only other confirmed records are of three individuals collected from the River Tamar near Gunnislake, South Devon. The Orange-spotted Emerald favoured deep, slow-moving sections of tree-lined rivers. Searches in apparently suitable sites have been made since the species' presumed extinction but have proved fruitless.

APPENDIX
Vice-Counties of Britain and Ireland

British Isles

1. West Cornwall (with Scilly)
2. East Cornwall
3. South Devon
4. North Devon
5. South Somerset
6. North Somerset
7. North Wiltshire
8. South Wiltshire
9. Dorset
10. Isle of Wight
11. South Hampshire
12. North Hampshire
13. West Sussex
14. East Sussex
15. East Kent
16. West Kent
17. Surrey
18. South Essex
19. North Essex
20. Hertfordshire
21. Middlesex
22. Berkshire
23. Oxfordshire
24. Buckinghamshire
25. East Suffolk
26. West Suffolk
27. East Norfolk
28. West Norfolk
29. Cambridgeshire
30. Bedfordshire
31. Huntingdonshire
32. Northamptonshire
33. East Gloucestershire
34. West Gloucestershire
35. Monmouthshire
36. Herefordshire
37. Worcestershire
38. Warwickshire
39. Staffordshire
40. Shropshire (Salop)
41. Glamorgan
42. Breconshire
43. Radnorshire
44. Carmarthenshire
45. Pembrokeshire
46. Cardiganshire
47. Montgomeryshire
48. Merionethshire
49. Caernarvonshire
50. Denbighshire

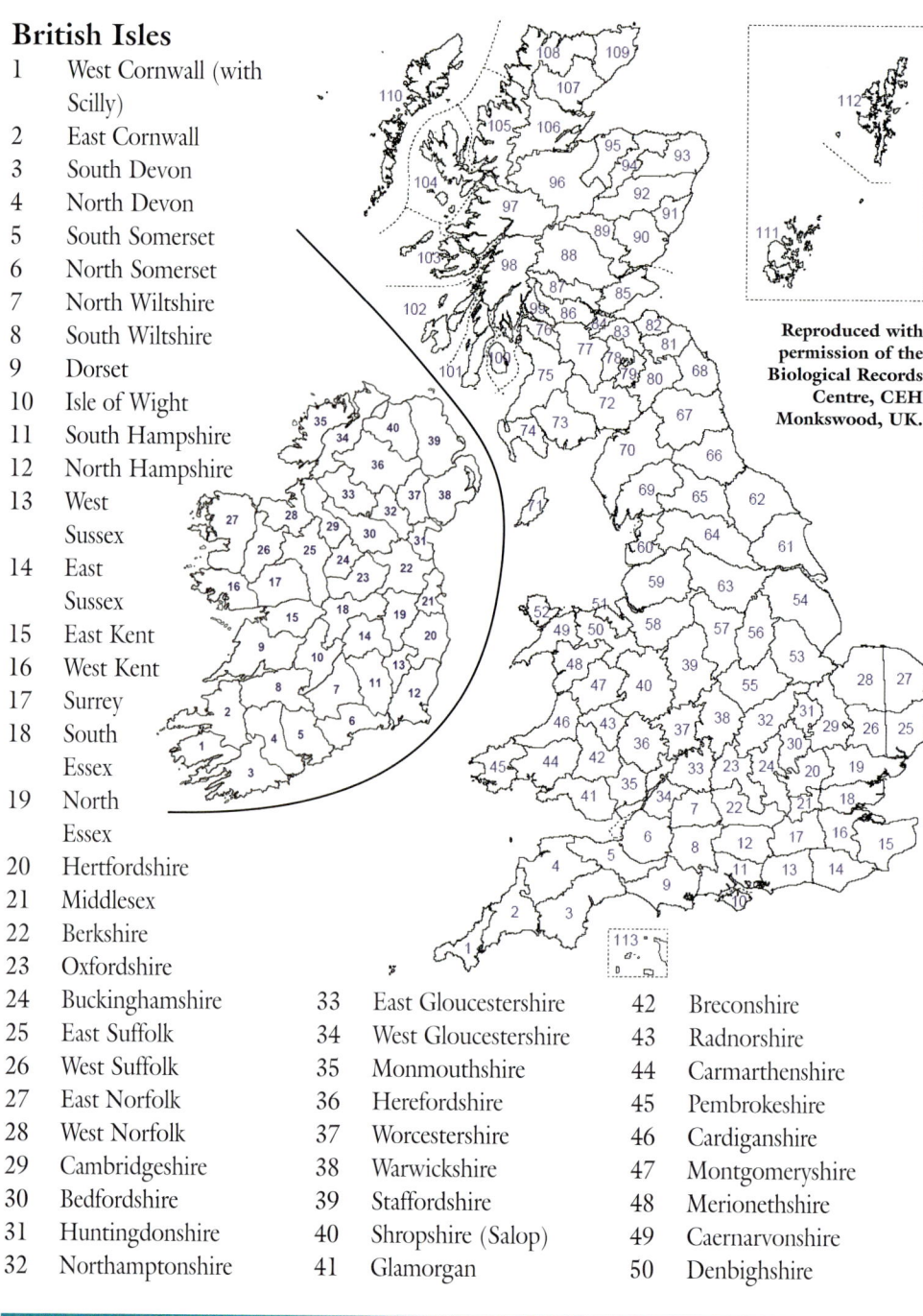

Reproduced with permission of the Biological Records Centre, CEH Monkswood, UK.

Vice-Counties of Britain and Ireland

51	Flintshire	72	Dumfriesshire	92	South Aberdeenshire
52	Anglesey	73	Kirkcudbrightshire	93	North Aberdeenshire
53	South Lincolnshire	74	Wigtownshire	94	Banffshire
54	North Lincolnshire	75	Ayrshire	95	Moray (Elgin)
55	Leicestershire (with Rutland)	76	Renfrewshire	96	East Inverness-shire (with Nairn)
56	Nottinghamshire	77	Lanarkshire	97	West Inverness-shire
57	Derbyshire	78	Peeblesshire	98	Argyll Main
58	Cheshire	79	Selkirkshire	99	Dunbartonshire
59	South Lancashire	80	Roxburghshire	100	Clyde Isles
60	West Lancashire	81	Berwickshire	101	Kintyre
61	South-east Yorkshire	82	East Lothian (Haddington)	102	South Ebudes
62	North-east Yorkshire	83	Midlothian (Edinburgh)	103	Mid Ebudes
63	South-west Yorkshire			104	North Ebudes
64	Mid-west Yorkshire	84	West Lothian (Linlithgow)	105	West Ross
65	North-west Yorkshire			106	East Ross
66	Durham	85	Fifeshire (with Kinross)	107	East Sutherland
67	South Northumberland	86	Stirlingshire	108	West Sutherland
68	North Northumberland (Cheviot)	87	West Perthshire (with Clackmannan)	109	Caithness
				110	Outer Hebrides
69	Westmorland with North Lancashire	88	Mid Perthshire	111	Orkney Islands
		89	East Perthshire	112	Shetland Islands (Zetland)
70	Cumberland	90	Angus (Forfar)		
71	Isle of Man	91	Kincardineshire	113	Channel Isles

Ireland

1	South Kerry	14	Leix	28	Sligo
2	North Kerry	15	South-east Galway	29	Leitrim
3	West Cork	16	West Galway	30	Cavan
4	Mid Cork	17	North-east Galway	31	Louth
5	East Cork	18	Offaly	32	Monaghan
6	Waterford	19	Kildare	33	Fermanagh
7	South Tipperary	20	Wicklow	34	East Donegal
8	Limerick	21	Dublin	35	West Donegal
9	Clare	22	Meath	36	Tyrone
10	North Tipperary	23	West Meath	37	Armagh
11	Kilkenny	24	Longford	38	Down
12	Wexford	25	Roscommon	39	Antrim
13	Carlow	26	East Mayo	40	Londonderry
		27	West Mayo		

References and further Reading

Askew, R. R. (2004) *The Dragonflies of Europe* (revised edition). Harley, Essex.

Birdguides (2006) *A DVD Guide to the Dragonflies and Damselflies of Great Britain*. Birdguides Ltd, Yorkshire.

Brooks, S. (2004) *Field Guide to the Dragonflies and Damselflies of Great Britain and Ireland*. British Wildlife Publishing, Hampshire.

Brooks, S., Cham, S. & Lewington, R. (2014) *Field Guide to the Dragonflies and Damselflies of Great Britain and Ireland*. British Wildlife Publishing, Hampshire.

Brownett, A. (2005) A re-examination of the status of the Norfolk Damselfly *Coenagrion armatum* (Charpentier): a species of Odonata now presumed extinct in Britain. *J. Brit. Dragonfly Soc.* **21**: 21-26.

Brullé, A. (1832) *Expédition scientifique de Morée*. **3**. Levrault, Paris.

Burmeister, H. C. C. (1839) *Handbuch der Entomologie*. **2**. Eslin, Berlin.

Campion, F. W. & Campion, H. (1905) On the Dark Form of *Ischnura elegans* (female). *Entom.* **38**: 298-299.

Cham, S., Nelson, B., Parr, A., Prentice, S., Smallshire, D. & Taylor, P. (2014). *Atlas of Dragonflies in Britain and Ireland*. Biological Records Centre, Shropshire.

Charpentier, T. de (1825) *Horae entomologicae*. Gosohorsky, Bratislavia.

Charpentier, T. de (1840) *Libellulinae europaea descriptae ac depictae*. Leopold Voss, Leipzig.

Corbet, P. S. (2000) The first recorded arrival of *Anax junius*, Drury (Anisoptera: Aeshnidae) in Europe: a scientist's perspective. *Int. J. Odonatology* **3**: 153-162.

Corbet, P. & Brooks, S. (2008) *Dragonflies*. New Naturalist Library, Collins, London.

Dale, J. C. (1834) *Cordulia curtisii* Dale, a new species hitherto undescribed, characterised by Mr. Dale. *Mag. nat. Hist.***7**: 60-61.

Dijkstra, K.-D. B. (2006) *Field Guide to the Dragonflies of Britain and Europe*. British Wildlife Publishing, Dorset.

Donovan, E. (1807) *The natural history of British Insects*. **12**. Rivington, London.

Drury, D. (1770) *Illustrations of Natural History; wherein are exhibited two hundred and forty forms of exotic insects*. **1**. White, London.

Dudley, S., Dudley, C. & Mackay, A. (2007) *Watching British Dragonflies*. Subbuteo, Shropshire.

Fabricius, J. C. (1798) *Supplementum Entomologiae systematicae*. Proft & Storch, Hafniae.

Hansemann, J. W. A. (1823) Anfang einer Auseinandersetzung der Deutschen Arten der Gattung Agrion. *Weidmann's zool. Mag.* **2**: 148-161.

Harris, M. (1782) *An Exposition of British Insects*. White & Robson, London.

Hill, P. & Twist, C. (1996) *Butterflies and Dragonflies: A Site Guide*. Arlequin, Essex.

Hill, P. & Twist, C. (1998) *Butterflies*

and Dragonflies: A Site Guide (second edition). Arlequin, Essex.

Jones, S. P. (1996) The first British record of the Scarlet Dragonfly *Crocothemis erythraea* (Brullé). *J. Brit. Dragonfly Soc.* **12**: 11-12.

Killington, F. J. (1924) *Ischnura elegans*, Lind.: Its teneral colour phases, and its mature varieties and aberrations. *Entomologist* **57**: 1-6.

Kirby, (1890) *A Synonymic catalogue of Neuroptera Odonata, or Dragonflies: with an appendix of fossil species*. Gurney & Jackson, London.

Latreille, P. A. (1805) *Histoire naturelle, générale et particullière, des Crustacaés et des Insectes*. **13**. Dufart, Paris.

Leach, W. E. (1815) Entomology. *Brewster's Edinburgh Encycl.* **9**: 57-172.

Linnaeus, C. (1758) *Systema Naturae* **1** (10th edition). Stockholm.

Lucas, W. J. (1900) *British Dragonflies*. Upcott Gill, London.

Lucas, W. J. (1912) British Odonata. *Entomologist* **45**: 171.

McGeeney, A. (1986) *A Complete Guide to British Dragonflies*. Cape, London.

Müller, O. F. (1764) *Fauna Insectorum friedrichsdalina*. Gleditisch, Hafinae & Lipsiae.

Müller, O. F. (1767) Enumeratio ac descriptio Libellularum agri Friedrichsdalensis. *Nova Acta Acad. Caesar. Leop. Carol.* **3**: 122-131.

Müller, O. F. (1776) (In: Allioni, C. (1766)) Manipulus insectorum Taurinensium. *Mél. Philos. Math. Soc. r. Turin.* **3**: 185-198.

Newman, E. (1833) Entomological notes. *Entomologist's Magazine* **1**: 505-514.

Pallas, P. S. (1771) *Reise durch verschiedene Provinzen des Russischen Reiches in den Jahren 1768-1774*. **1**. Petersburg.

Parkes, K. A., Amos, W., Moore, N. W., Hoffmann, J. I. & Moore, J. (2009) Population structure and speciation in the dragonfly Sympetrum striolatum/nigrescens (Odonata: Libellulidae): an analysis using AFLP markers. *European Journal of Entomology* **106**: 1279-1284.

Parr, A. J. (2009) Winter Damselfly *Sympecma fusca* Vander Linden in West Glamorgan. *Atropos* **37**: 28-31.

Parr, A. J. (2010) Records of Exotic Odonata in Britain during 2010. *Atropos* **41**: 39-42.

Pellow, K. (1999) An influx of Green Darner *Anax junius* (Drury) into Cornwall and the Isles of Scilly: the first European records. *Atropos* **6**: 3-7.

Pilgrim, E. M. & Von Dohlen, C. D. (2007) Molecular and morphological study of species-level questions within the dragonfly genus Sympetrum (Odonata: Libellulidae). *Annals of the Entomological Society of America* **100**: 688-702.

Rambur, J. P. (1842) *Histoire naturelle des insectes. Néuroptères*. Roret, Paris.

Riley, A. M. & Prior, G. (2003) *British*

Refernces and further Reading

and Irish Pug Moths. Harley, Colchester.

Rose, F. (2006) *The Wild Flower Key*. Warne, London.

Sélys Longchamps, E de, (1837) *Catalogue des Lépidoptères ou papillons de Belgique, précédé du tableau des Libellulines de ce pays*. Liege.

Sélys Longchamps, E de, (1839) Description de deux nouvelle espèces d'Aeshna du sous-genre *Anax*. *Bull. Acad. r. Belg*. **6**: 386-393.

Sélys Longchamps, E de, (1840) *Monographie des Libellulidées d'Europe*. Paris & Brussels.

Sélys Longchamps, E de, (1876) Synopsis des Arionines, 5e légion: *Agrion* (suite). *Bull. Acad. R. Belg*. **41** (second series): 1-282.

Sherwin, G. (2010) *British Damselflies* (DVD). Privately published.

Sherwin, G. (2011) *British Dragonflies* (DVD). Privately published.

Smallshire, D. & Swash, A. (2004) *Britain's Dragonflies: A Field Guide to the Damselflies and Dragonflies of Britain and Ireland* (first edition). WildGuides, Hampshire.

Smallshire, D. & Swash, A. (2010) *Britain's Dragonflies: A Field Guide to the Damselflies and Dragonflies of Britain and Ireland* (second (revised) edition). WildGuides, Hampshire.

Smallshire, D. & Swash, A. (2014) *Britain's Dragonflies: A Field Guide to the Damselflies and Dragonflies of Britain and Ireland* (third (revised) edition). WildGuides, Hampshire.

Smallshire, D. & Swash, A. (2018) *Britain's Dragonflies: A Field Guide to the Damselflies and Dragonflies of Britain and Ireland* (fourth (revised) edition). WildGuides, Hampshire.

Stephens, J. F. (1835) *Illustrations of British Entomology*. Mandibulata. **6**: 35. Baldwin & Cradock, London.

Ström, H. (1783) *Kort underretning om Eger Sognekald. Nye Saml. K. Danske Vidensk. Selsk skr*. **2**: 569-580.

Sulzer, J. H. (1776) *Abgekürzte Geschichte der Insekten nach dem Linneischen System*. Steiner, Winterthur

Thompson, R. & Nelson, B. (2014) *Guide to the Dragonflies and Damselflies of Ireland*. National Museums Service of Northern Ireland. Cultra.

Vander Linden, P. L. (1820) *Agriones bononiensis descripta*. Bononiae

Vander Linden, P. L. (1823) Aeshnae bononienses descriptae, adjecta annotatione ad Agriones bononienses descriptas. *Opusc. Scient*. **4**: 158-165.

Vander Linden, P. L. (1825) *Monographiae Libellulinarum Europaearum Specimen*. Frank, Brussels.

Villers, C. J. (1789) *Caroli Linnaei Entomologica*. **3**. Piestre & Delamollière, Lugduni.

Zetterstedt, J. W. (1840) *Insecta Lapponica descripta*. Voss, Lipsiae.

Scarce Emerald Damselfly

Index

Damselfly Group 9 41, 43, 45, 47, 49, 51, 65, **85-95**, 231

Damselfly Group 10 49, 55, 65, **96-8**

Damselfly Group 11 39, 62, 65, **98-100**

Damselfly Group 12 51, 57, 60, 65, **101-3**

Damselfly Group 13 41, 43, 45, 47, 53, 57, 60, 65, **104-12**

Denbighshire 146, 238

Dersingham Bog, West Norfolk 161

Devon 32, 38, 44, 52, 58, 119, 138, 160, 168, 234, 237-8

Donegal 154, 239

Dorset 36, 38, 44, 58, 131, 138, 160, 237, 238

Down 42, 157, 168, 239

Downy Emerald Dragonfly 19, 113, **138-40**, 143, 145, 189, 191-5

Dragonfly Group 1 121, 128, 170, **172-3**

Dragonfly Group 2 138, 170, **174**

Dragonfly Group 3 115, 118, 122, 126, 130, 133, 170, **174-82**, 233

Dragonfly Group 4 116, 118, 122, 126, 130, 133, 170, **183-7**, 233

Dragonfly Group 5 136, 170, **188**

Dragonfly Group 6 140, 143, 145, 170, **189-95**

Dragonfly Group 7 149, 153, 170, **196-7**

Dragonfly Group 8 149, 151, 156, 159, 171, **197-203**

Dragonfly Group 9 149, 151, 153, 156, 171, **204-7**

Dragonfly Group 10 164, 167, 169, 171, **208-11**

Dragonfly Group 11 159, 162, 164, 167, 169, 171, **212-25**

Dragonfly Group 12 147, 171, **225**

Dragonfly Group 13 147, 162, 171, **226-7**

Dragonfly Group 14 151, 156, 159, 167, 187, **228-9**

Dropwing 157

Dublin 34, 239

Dumfriesshire 48, 129, 131, 239

Dunbarton 141, 239

Durham 34, 48, 239

East Anglia 36, 121, 161

Ebudes 34, 239

Elder 27

Emerald Damselfly 16, 18, 25, 27-9, **30-1**, 68-75

Emperor Dragonfly 19, 113, 118-9, 122, 126, **129-31**, 133, 140, 153, 174-5, 177-8, 180-4, 186-7

Enallagma cyathigerum (Charpentier) 18, 25, **50**, 93, 101

erythrogastrum Sélys, *Ceriagrion tenellum* f. 18, 39

Erythromma najas (Hansemann) 18, 25, **52**, 80, 111

Erythromma viridulum (Charpentier) 18, 25, **54**, 80, 97

Esox lucius Linn. 42

Essex 28, 36, 124, 230-2, 238

Fakenham, West Norfolk 156

Common Darter

Index

Fenn's Moss, Shropshire 146

Fermanagh 42, 168, 239

Fibia. See Anatomy

Fifeshire 154, 239

Foulshaw Moss, Cumbria 146

Four-spotted Chaser 19, 113, 131, 149, 151, **152-3**, 156, 196-7, 204, 206-7

Frons. See Anatomy

fulvipes Stephens, *Pyrrhosoma nymphula* f. 18, 62

Galway 28, 34, 139, 154, 239

Glamorgan 44, 54, 139, 235, 238

Glen Gour, West Inverness-shire 142

Gloucestershire 138, 150, 233, 238

Golden-ringed Dragonfly 10, 19, 118, **137-8**, 174

Gomphus flavipes, see *Stylurus flavipes*

Gomphus vulgatissimus (Linn.) 19, 113, **134**, 188

Goring, Oxfordshire 134-5

Green Darner 20, 235

Hairy Dragonfly 19, 113, 118-9, 122, 126, 130, **131-3**, 174-5, 177-9, 181-4, 186-7

Hairy Hawker, see Hairy Dragonfly

Hampshire 34, 36, 38, 44, 58, 144, 150, 157, 160, 238

Hawthorn 45, 53

Hemianax ephippiger, see *Anax ephippiger*

Highland Darter 19, 167, 208, 212, 214-5

Holkham Meals, West Norfolk 125

Holt Country Park, West Norfolk 120

Holt Lowes, West Norfolk 158

Hornwort 54

Inferior anal appendage. See Anatomy

Inferior claspers. See Anatomy

infuscans Campion & Campion, *Ischnura elegans* f. 18, 41, 43, 47, 53, **57**, 60, 104-10, 112

infuscans-obsoleta Killington, *Ischnura elegans* f. 18, 51, **57**, 101-3

intermedium Sélys, *Ceriagrion tenellum* f. 18, 39

Intersegmental divisions. See Anatomy

Inverness-shire 34, 40, 131, 139, 144, 234

Irish Damselfly 18, 25, **42-3**, 47, 49, 51, 53, 57, 60, 85, 87, 89-92, 94, 104-11

Ischnura elegans (Vander Linden) 18, 25, **56**, 82, 103, 108

Ischnura pumilio (Charpentier) 18, 25, **58**, 84, 101, 187

Ischnura senegalensis (Rambur) 20, 235

Island of Muck 141

Isle of Man 26, 28, 32, 34, 36, 38, 44, 46, 48, 52, 54, 58, 116, 119, 121, 124, 129, 131, 134, 137, 139, 141, 144, 146, 148, 150, 154, 157, 160, 163, 168, 239

Isle of Mull 141, 167, 174, 196, 208, 215

Isle of Wight 157, 238

Isles of Scilly 26, 28, 30, 32, 34, 36, 38, 44, 46, 48, 52, 54, 58, 61, 116, 119, 121, 124, 129, 131, 134, 137, 141, 144, 146, 148, 150, 152, 154, 157, 160, 163, 166, 234-6, 238

Index

isoseles, see *Anaciaeshna isosceles* (Müller)

Kelling Water Meadows, East Norfolk 163

Kent 28, 34, 36, 127, 230-2, 235-6, 238

Kerry 34, 139, 141, 239

Kincardineshire 166, 239

Kinlochewe, West Ross 141

Kintyre 116, 131, 239

Kirkcudbrightshire 48, 114, 129, 131, 157, 239

lactea (Charpentier), *Platycnemis pennipes* f. 18, 37, 76

Lake District 116, 129

Lancashire 32, 54, 119, 232, 236, 239

Large Red Damselfly 18, 25, 39, **61-2**, 77-9, 98-100

Large White-faced Darter 20, 236

Leitrim 42, 239

Lesser Emperor Dragonfly 20, 239

Lestes barbarus (Fabricius) 19, 230

Lestes dryas Kirby 18, 25, **28**, 68, 73

Lestes sponsa (Hansemann) 18, 25, **30**, 68, 73

Leucorrhinia dubia (Vander Linden) 19, 113, **146**, 225-6

Leucorrhinia pectoralis (Charpentier) 20, 236

Libellula depressa Linn. 19, 113, **148**, 196, 198, 204

Libellula fulva Müller 19, 113, **150**, 197, 204, 228

Libellula quadrimaculata Linn. 19, 113, **152**, 196, 207

Limerick 125, 154, 239

Lincolnshire 48, 127, 232, 239

Loch Bran, East Inverness-shire 145

Longmynd, Shropshire 30, 123

Louth 154, 239

melanogastrum Sélys, *Ceriagrion tenellum* f. 18, **39**, 62, 98-100

melanotum Sélys, *Pyrrhosoma nymphula* f. 18, **39**, 62, 98-100

Mayo 157, 239

Merionethshire 139, 238

Middlesex 28, 238

Midlothian 148, 239

Migrant Hawker 19, 113, 118-9, 222, **124-6**, 130, 133, 174-5, 177-87, 232-3

Mill Lawn Brook, Hampshire 35, 38, 45

Monaghan 42, 239

Montgomeryshire 139, 238

Montiaghs Moss, Antrim 42-3

Moorland Hawker 121

Moray 116, 141, 239

Myrica gale 159

New Forest, South Hampshire 30, 34-5, 37-9, 44, 50, 52, 58, 61, 129, 131, 137, 148, 157

nigrescens Lucas, *Sympetrum striolatum* f. 19, 159, 162, 164-5, 167, 169, 208, 214-6

Norfolk 7, 10-1, 16, 28, 32, 38, 58, 116, 127, 138, 150, 157, 160, 230, 232, 236-7, 238

Norfolk Damselfly 18, 25, 40-1, 47, 49, 51, 53, 57, 60, 85-6, 88-92, 94, 104, 106-11

Index

Norfolk Hawker 19, 113, 121, **127-8**, 172-3

Northamptonshire 138, 238

Northern Damselfly 18, 25, **40-1**, 47, 49, 51, 53, 57, 60, 85-6, 88-92, 94, 104, 106-11

Northern Emerald Dragonfly 19, 113, 140, **141-3**, 145, 189-95

Northumberland 32, 54, 129, 148, 239

Ober Water, Hampshire 37

Ocular spots. See Anatomy

Orange-spotted Emerald Dragonfly 20, 237

Orkney Islands 137, 166, 233-4, 239

Orthetrum cancellatum (Linn.) 19, 113, **154**, 201, 206, 228

Outer Hebrides 50, 137, 157, 239

Oxygastra curtisi (Dale) 20, 237

Oxygastra curtisii, see *Oxygastra curtisi* (Dale)

Pantala flavescens (Fabricius) 20, 236

Paxton Pits, Cambridgeshire 127

Pembrokeshire 44, 238

Pennines 116, 129

Pensthorpe, West Norfolk 13, 53

Perthshire 40, 239

Pike 42

Pingo 28, 118

Post-ocular bar. See Anatomy

Platycnemis pennipes (Pallas) 18, 25, **36-7**, 76

praenubila Newman, *Libellula quadrimaculata* f. 19, **153**, 196-7

Pseudo-pterostigmata. See Anatomy

Pterostigmata. See Anatomy

Pulborough, West Sussex 151

Pyrrhosoma nymphula (Sulzer) 18, 25, **61**, 78, 98

Red-eyed Damselfly 18, 25, 47, **52-3**, 55, 57, 60, 80-2, 104, 107, 109, 110-1

Red-veined Darter 19, 113, 159, 162, **163-5**, 167, 169, 208-12, 214, 216-8, 221-2, 224-5

Reedham, East Norfolk 56

River Arun 134, 144

River Dee 134

River Ouse 144

River Severn 134

River Teifi 134

River Thames 134-5

River Twyi 134

River Wensum 32

River Wye 134

Roxburghshire 154, 239

Roydon Common, West Norfolk 160

Rudd 42

Ruddy Darter 19, 113, 159, 162, 164-5, 167, **168-9**, 208, 210-3, 215-6, 218-9, 221-4

rufescens Stephens, *Ischnura elegans* f. 18, 57, 82-4

rufescens-obsoleta. See *infuscans-obsoleta* Killington

Salix 26-7

Sambucus nigra 27

Scaleby Moss, Cumbria 146

Golden-ringed Dragonfly

Index

Scarce Blue-tailed Damselfly 18, 25, 45, 47, 51, 53, 57, **58-60**, 82-4, 101-4, 106-7, 109-12

Scarce Chaser 7, 19, 113, 149, **150-2**, 156, 159, 167, 197--207, 228

Scarce Emerald Damselfly 16, 18, 25, 27, **28-9**, 31, 68-75

Scardinius erythrophthalmus (Linn.) 42

Scarlet Darter 20, 236

Sculthorpe, West Norfolk 13, 32-3, 54-5, 132

Shetland 30, 50, 56, 121, 137, 160, 166, 234, 239

Shropshire 30, 36, 48, 138, 146, 150, 238

Small Red Damselfly *Ceriagrion tenellum* (Villers) 18, 25, **38-9**, 62, 77-9, 98-100

Small Red-eyed Damselfly 18, 25, 47, 49, 52, **54-5**, 80-2, 96-8

Snowdonia 116

Somatochlora arctica (Zetterstedt) 19, 113, **140**, 194

Somatochlora flavomaculata (Vander Linden) 20, 235

Somatochlora metallica (Vander Linden) 19, 113, **144**, 192

Somerset 36, 48, 131, 138, 157, 230, 238

Southern Damselfly 18, 25, **44-5**, 47, 49, 51, 53, 57, 60, 85, 88-93, 104-5, 107-11

Southern Emerald Damselfly 19, 70, **230**

Southern Hawker 18, 113, **115-9**, 122, 126, 130, 133, 174, 176, 178-87

Southern Migrant Hawker 20, 174, 189, 230, **231-2**

Sphagnum 114, 161

Sparham Pools, Norfolk 47

Staffordshire 138, 146, 238

Stratiotes aloides 127

Strumpshaw Fen, East Norfolk 7, 48, 127, 150, 165

Stylurus flavipes (Charpentier) 235

Superior claspers. See Anatomy

Suffolk 127, 235-6, 238

Surrey 28, 38, 144, 157, 160, 238

Sussex 121, 144, 160, 235, 238

Sympecma fusca (Vander Linden) 20, 235

Sympetrum danae (Sulzer) 10, 19, 113, **160**, 220, 226

Sympetrum flaveolum (Linn.) 20, 236

Sympetrum fonscolombii (Sélys) 19, 113, **163**, 209, 217

Sympetrum nigrescens Lucas 167

Sympetrum pedemontanum ([Müller]) 20, 236

Sympetrum sanguineum (Müller) 19, 113, **168**, 210, 222

Sympetrum striolatum (Charpentier) 19, 113, **165**, 208, 214, 229

Sympetrum vulgatum (Linn.) 20, 236

tenellum (Villers), *Ceriagrion tenellum* f. 38

Thompson Common, West Norfolk 28, 117-8, 168

Thompson Water, East Norfolk 152, 155

Tibia. See Anatomy

Titchwell, West Norfolk 26

Index

Upton Fen, East Norfolk 46, 131

Vagrant Darter 20, 236

Vagrant Emperor Dragonfly 20, 239

Variable Damselfly 18, 25, 43, 47, **48-9**, 51, 55, 85, 87, 89-92, 94-8

violacea Sélys *Ischnura elegans* f. 18-9, **56**, 57, 82-4

Vulvar scale. See Anatomy

Wandering Glider 20, 236

Wat Tyler Country Park, South Essex 124

Water-soldier 127

Waterford 154, 239

West Cork 141, 157, 239

West Lothian 119, 239

West Ross 114, 141, 146, 157, 239

Westbere Lakes, West Kent 127

Westmorland 34, 138, 157, 239

White-faced Darter 19, 113, **146-7**, 162, 225-6

White-legged Damselfly 18, 25, **36-7**, 76

Whixall Moss, Shropshire 146

Wicken Fen, Cambridgeshire 48

Wicklow 157, 239

Wigtownshire 48, 129, 131, 239

Willow 26-7

Willow Emerald Damselfly 18, 25, **26-7**, 29, 31, 70-5

Wiltshire 34, 36, 150, 238

Winter Damselfly 20, 235

Winterton, East Norfolk 59, 230

Worcestershire 150, 239

Yellow-legged Clubtail 20, 235

Yellow-spotted Emerald 20, 235-6

Yellow-winged Darter 20, 235

Yorkshire 26, 34, 52, 119, 121, 131, 157, 160, 232, 239

Beautiful Demoiselle

Other Nature Books by Brambleby Books

 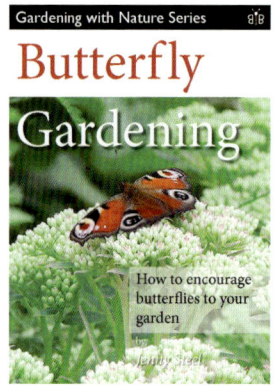 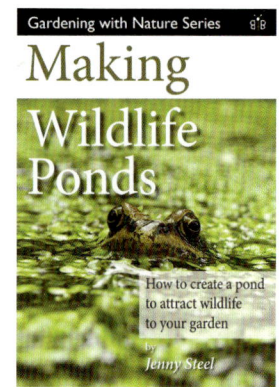

Making Garden Meadows - How to create a natural haven for wildlife
Jenny Steel
9781908241221

Butterfly Gardening - How to encourage butterflies to your garden
Jenny Steel
9781908241436

Making Wildlife Ponds - How to create a pond to attract wildlife to your garden
Jenny Steel
9781908241481

What's in your Garden - A book for young explorers
Colin Spedding
9780955392818

Buzzing! Discover the poetry in garden minibeasts
Anneliese Emmans Dean
9781908241443

Flying High! Discover the poetry in British birds
Anneliese Emmans Dean
9781908241504

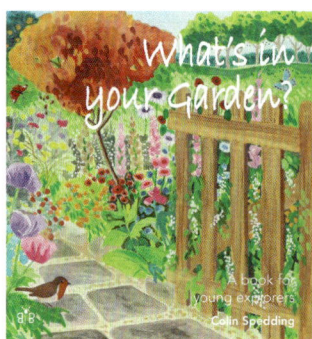 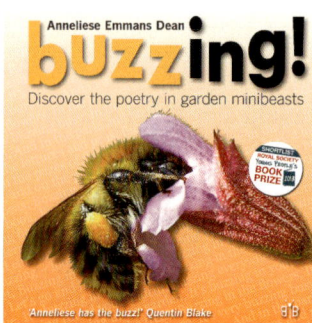 

Other Nature Books by Brambleby Books

 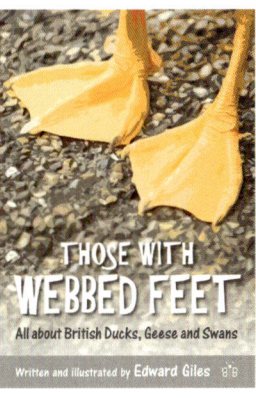 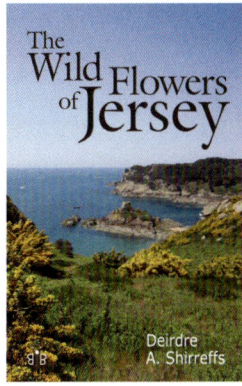 

The Wild Flowers of The Isle of Purbeck, Second Edition
Edward A. Pratt
9781908241450

Those with Webbed Feet - All about British Ducks, Geese and Swans
Edward Giles
9781908241573

The Wild Flowers of Jersey
Deirdre A. Shirreffs
978 1908241 337

British and Irish Butterflies - The complete Identification, Field and Site Guide
Adrian M. Riley
9780955392801

The elephant and the mountain
Richard ffrench-Constant
978 1908241 528

The boy and the trout
Richard ffrench-Constant
978 1908241 580

Zoooo... Living Poems for Children
Hugh David Loxdale
9781908241139

 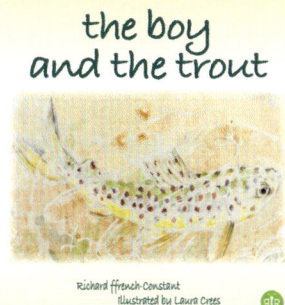 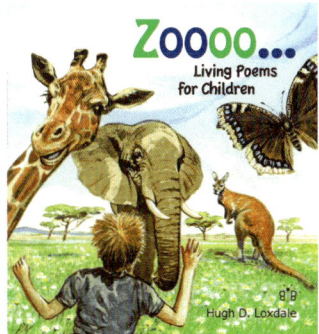

Brambleby Books

www.bramblebybooks.co.uk